P9-BHR-844

The Earthsteward's Handbook

The Sevenfold Path of Peace

This book was originally developed for Holyearth Foundation and the Earthstewards Network. The premise of the Earthstewards Network is that if we create and work toward positive alternatives, we can promote attitudes of faith, trust and effectiveness in the world and thus use the massive energies of change of our times for positive transformation.

Written and compiled by
Danaan Parry and Lila Forest

Illustrations by
Susan Hand

Cover Photograph by
Linda Joy Montgomery

Graphic Design by
Doreen DeNicola

Printed on Recycled Paper

Published by
Sunstone Publications
P.O. Box 788E Cooperstown, NY 13326

Library of Congress Catalog Number: 88-134579
ISBN 0-913319-19-8

Preface

This book has been created to provide a vehicle for bringing spirit into matter; that is, it is a collection of reflections, ideas and practical suggestions to help make the spiritual vision of the Sevenfold Path of Peace a reality in our life together on this planet.

There are nine sections to the text: one introductory in nature, one for each of the aspects of the Sevenfold Path, and an appendix listing organizations whose work contributes to planetary transformation. Each of the seven sections contains articles to deepen awareness in that area, and each contains a chapter entitled, "What You Can Do," which gives many practical ideas for making that aspect more fully manifest in daily life.

Use this book to stimulate your own creative process in thinking and feeling about these areas and in finding ways to implement them in your personal life and in the spheres in which you have influence. Write to the Earthstewards Network and tell us about other thoughts and ideas not mentioned here which might be useful to our readers and to members of the Earthstewards Network (see the article with that title in the Introductory Section). This book is by no means comprehensive; the ideas and contacts given are only a beginning.

With this book go our best wishes to you and the hope that it will be a blessing for your journey.

Danaan Parry
For the Earthstewards Network

Acknowledgments

We wish to thank all the people who have kindly contributed their work to this book. Copyright credit is given for the quotes reprinted from THE PROPHET, by Kahlil Gibran, by permission of Alfred A. Knopf, Inc. Copyright 1923 by Kahlil Gibran and renewed 1951 by Administrators C.T.A. of the Kahlil Gibran Estate and Mary G. Gibran. We also wish to thank the Theosophical University Press for permission to quote excerpts from the 1979 3rd and Revised Edition of GOLDEN PRECEPTS OF ESOTERICISM by G. de Purucker. The quote from GITANJALI by Rabindranath Tagore is used with permission by Branden Publishing Company, Boston Massachusetts.

We have made many attempts to contact all contributors and have acknowledged them throughout the book. If anyone was inadvertently left out, please accept our apologies and let us know so you may be included in the next printing.

Weaving Together Our World Family With Love

In 1979, I sat on an idyllic mountain top in the redwood forest of Northern California, reflecting on the growth of the spiritual community that I had helped to create in this remote and beautiful place five years before. Since early 1975, we had lived here as a new breed of monks, following the Essene Way and attuning to the rhythms and cycles of the seasons, the moon and sun. There was peace in me, more than I had ever known.

And then I heard it, coming from deep in a valley miles away. The whine of a chain-saw cut through my reveries like it was sawing my own body. As I looked out over the mountain crests towards the sea, a thin blue haze hung above them, partially blocking the sun. Had this layer of pollution from the valley been here before? Had it been there each evening as I meditated, and I simply was oblivious to it? Had my peace-filled life-style be come a place to hide from reality?

For years, I had been training myself to listen to my still, small voice within, the one that always tells me truth, although it's usually what I don't want to hear. Now evoking that inner voice brought tears, and a recognition that my years on the mountain were through. I must now bring this peace, this connectedness to the earth and to life, down from the mountain top and into the cities, the freeways, the marketplace. I am reminded that 2000 years ago, the Essenes made that crucial decision to return from the peace of the desert to the cities, to prepare the way for the next step in consciousness.

In early 1980 Lila Forest and I created the Holyearth Foundation. We began writing the *Earthsteward's Handbook* as a tool, a guide for bringing back the awareness of the connectedness of all things and the sacredness of all life. We envisioned the Earthsteward's Network as the "body" of Holyearth Foundation, a loose-knit network of caring individuals who supported one another in the work of melting the wall so fear and isolation, and weaving together our world family with love.

In the *Earthsteward's Handbook* are the essences of the teachings and awarenesses we have gleaned from the Essenes, Taoism, Native American and other indigenous cultures. The thread that weaves them together is the Sevenfold Path of Peace, a mandala of interconnected steps for living in attunement with the earth and all its creatures.

Danaan Parry

The Sevenfold Path of Peace

In November of 1980, my son and I were driving up from Santa Cruz, and I was trying to beat the rush-hour traffic through San Francisco. And all of a sudden there it was, in the car, filling the space like a giant one-celled, glowing, living thing. It was a presence filled with love, impossible to ignore. I hit the brakes instinctively and somehow maneuvered the car over to the side of the freeway without sideswiping anyone. My son could feel it; I could feel and see it. I just sat there for awhile, looking at this softly glowing mandala. The image had seven groupings of works. I reached for my journal and recorded the words as they hovered there inside my car.

The result was the Sevenfold Path of Peace, which now serves as a major guideline for the Earthstewards Network. Where did it come from? It doesn't really matter to me. It represents an encompassing, wholistic model for my inner and outer growth, and that's what counts. When I take the time to focus on each of the seven areas and make them real in my life, I know that I am touching in to all the parts of me that need to be awakened. And because I am a holographic chip of the whole, my personal awakening is intimately connected to our common awakening. The Sevenfold Path of Peace has become a tool for my journey home. I'm honored to share that tool with you.

Danaan Parry

The Sevenfold Path of Peace

1

When we are at peace within our own hearts we shall be at peace with everyone and with our Mother the Earth.

2

When we recognize that our planet itself is a living organism coevolvong with humankind we shall become worthy of stewardship.

3

When we see ourselves as stewards of our planet & not as owners and masters of it there shall be lasting satisfaction from our labors.

4

When we accept the concept of Right Livelihood as the basic right of all we shall have respect for one another.

5

When we respect the sacredness of all life we shall be truly free.

6

When we free ourselves from our attachment to our ego-personalities we shall be able to experience our Oneness.

7

When we experience our Oneness—our total connectedness with all being, we shall be at peace within our own hearts.

Table of Contents

Table of Contents

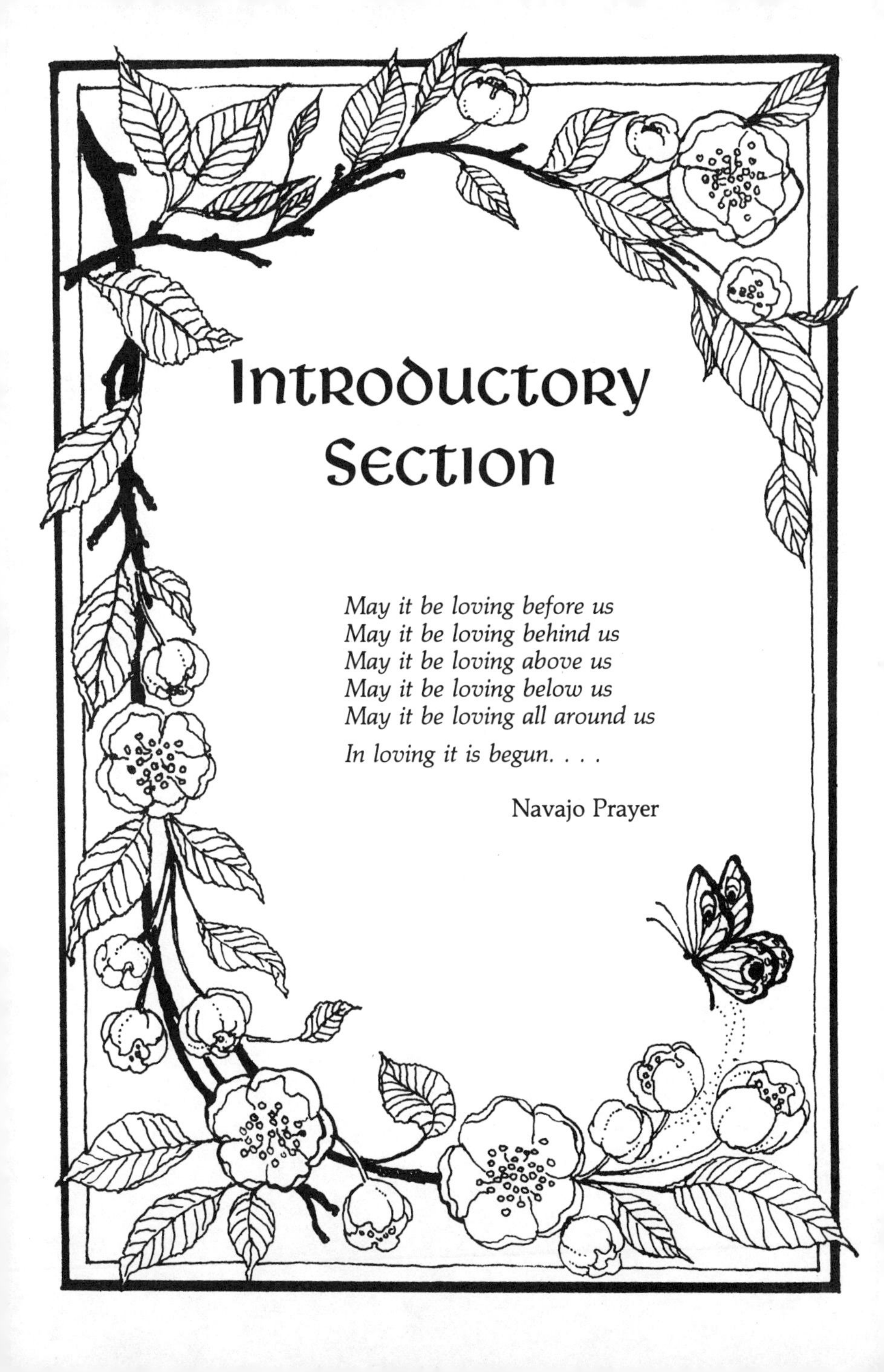

Introductory Section

May it be loving before us
May it be loving behind us
May it be loving above us
May it be loving below us
May it be loving all around us

In loving it is begun. . . .

Navajo Prayer

Toward a New Age

We see the signs of change all around us: explosions of consciousness, deeper spiritual awareness, agitation and conflict in the economic, political and social spheres, ecological crises, earthquakes, and psychic quakes. I personally believe we ain't seen nothin' yet, and that the years 1980–2000 will be a period of change and evolution that would stagger us, could we see it all in advance. In times of such rapid change, much of what we have held onto for security and safety thus far won't work for us any more. What will get us through, sustaining and nourishing us? Above all, we must ground in the Spirit; we must recognize that the root of all-that-is is our Oneness, the one force of life, love, light; and that the material plane is a shifting, changing, reflection of how well we see and know that one force. And from that one Source we can learn to apply three principles that make for a full, rich, joyous life, especially in times of great change. Let's explore these three in light of the present and coming times.

Letting Go

"Totally involved, totally unattached" (Ram Dass). If we could live in such a way, what joy would be ours! To be fully present to relationships, to the beauty of the world we live in, to the abundance of gifts given us, and to be able to let go of all these and be fully present to whatever is given us next; to live as a dance, letting our movements be one with the shifting rhythms of the cosmic music. In times of change, such an approach to life is priceless; it means the difference between a series of painful separations from what is known, and surrender to the flow of the river.

Attunement

In order to be able to respond to shifting energy patterns, and to be in the "right" place at the "right" time, we need to be in attunement with the earth, with the collective knowledge of humanity, and with the wisdom and guidance of spiritual powers. While it is futile to seek all of this with our conscious minds, we can be in touch with it on the intuitive levels if we use the tools available to us. Meditation, work with our dreams, and pendulum work are all very helpful for connecting our conscious awareness with our deeper selves, and for allowing that totally wise and loving part of ourselves to guide and enrich our lives more and more.

Attunement to each other for mutual support and working together comes with the regular practice of consciously joining our energies in circles for healing, chanting, meditation, and blessing our food and other gifts. The more we tune into each other in these ways, the more harmonious our life together becomes.

Cooperation

The crucial lesson and test of these times, as I see it, is that we must learn to act together in love and harmony, with full respect for the uniqueness of each one of us, or we must founder alone in isolation and loss. Getting in the habit of sharing what we have, recycling what we don't need, asking for and offering assistance, recognizing how what each does affects all others, working together, playing together, meditating and praying together – all these are aspects of cooperation. In our past human history we have experienced tribal oneness with little awareness of each member, and in recent times the exaltation of the individual with very low group consciousness. We have the opportunity now to find the balance between these two that nurtures each being and results in harmony of the whole. I know that this is possible as we move more in that direction in Holyearth and other groups in which I participate.

The Earthstewards Network

Planetary Transformation Follows Personal Transformation

The concept of the Earthsteward has emerged from a growing awareness of the intimate and total connection of all beings. This connectedness transcends all established lines of race, nationality, belief and even species. In this time of planetary crisis, an energy form is being tapped, which, if developed and released in a loving, compassionate manner, can utilize this time of crisis as a doorway to a new, more holistic existence. This energy form is our own human consciousness, elevated to the level at which we become clear about our real reasons for being here on this planet. When we, as Earthstewards, begin to accept responsibility for our part in the unfolding drama of the evolution of consciousness, then this transformation of our own personal lives affects all other life on our planet, and planetary transformation is possible. This is the essence of the Earthstewards Network: taking personal responsibility for one's part in it all. Earthstewards are people everywhere, connected by a network of communication and consciousness, who know the power of their thoughts and actions and are directing them in the service of their brothers and sisters and of their planet.

Earthstewards are agents of conscious, loving change who individually commit themselves to a program of service which they themselves have conceived. In small ways and in large ways, within themselves and within their sphere of influence, they facilitate the shift from a system based on competition, aggression, suspicion and intolerance to a new consciousness based on cooperation, sharing, openness and acceptance.

You are invited to become an Earthsteward. Fill out the application blank at the back of this book and mail it to the Earthstewards Network. You will be asked to make a commitment of your own choosing, conceived by you and actualized by you. It can change the world. Our present world of aggression, excess, and isolation will not be changed by massive counter-movements; it will be transformed by you and me and thousands like us who make a commitment, who take on a sacred obligation to make a difference in whatever ways we can, with the tools we now have.

You and the world stand on the brink of the next level of consciousness. For your own transformation—for our planet's transformation—take the leap. Become an Earthsteward!

On Commitment

We must alter our lives in order to alter our hearts, for it is impossible to live one way and pray another.

William Law

As we approach the millennium, we enter a time for drawing spirit into matter, so that our daily life is elevated to the sacred. How to accomplish this? Edgar Cayce said that patience, persistence, and consistency are spiritual qualities, and ancient Eastern advice on daily spiritual practice says:

When bored, do your practice.
When excited, do your practice.
When depressed, do your practice.
When joyful, do your practice.

The essence of commitment is in your ability to rise above ups and downs, the dance of the emotions, and to see the higher purpose for which the commitment was made. To make a commitment to yourself is to enter into a sacred trust with the highest part of you. There can be no more sacred commitment than that which affects, in a positive manner, the expansion of love and consciousness.

Kahlil Gibran wrote, "Your daily life is your temple and your religion. Whenever you enter it, take with you your all." Make your commitments consciously and use them as vehicles for taking your all with you wherever you go.

INNER PEACE

"Peace I ask of thee, O River,
Peace, peace, peace.
When I learn to live serenely,
Cares will cease.
From the hills I gather courage,
Vision of a day to be,
Strength to lead and faith to follow,
All are given unto me.
Peace I ask of thee, O River,
Peace, peace, peace."

—Traditional Song

WHEN WE ARE AT PEACE
WITHIN OUR OWN HEARTS
WE SHALL BE AT PEACE
WITH EVERYONE
AND WITH
OUR MOTHER
THE EARTH

Peace Within

The first statement of the Sevenfold Path of Peace speaks of the beginning point of any conscious journey. It says, "When we are at peace within our own hearts, we shall be at peace with everyone and with our Mother the Earth."

The path to anywhere meaningful always begins within one's own heart. And if anyone should desire to be of service, to make a difference, to be a channel for peace or love or kindness in the world, then first that person must be at peace within. First that person must love her/himself.

Inner peace is not a state reserved for monks and yogis; it is available to each of us, and it is one of the nicest gifts that we can give to ourselves. Because it is a feeling state, and not a thinking state, it is difficult to describe in words. Perhaps it can be seen as a harmony of body, mind and spirit, a quieting of the mind-chatter that drones on and on in our heads when we are not at peace with ourselves. It is a feeling that one is in this world but not of it, not hooked into the somber seriousness of the melodrama. Some see it as being in touch with one's compassionate witness, which is that part of you that can stand back and watch "you" do your dance of daily life, and love it all, and love you, and love even that about it and you that "you" think is unlovable. One thing is sure: inner peace is found not through doing, but through the surrender of doing. And this is perhaps the most difficult step. We are so conditioned to try to achieve and, if it isn't working, to try harder. These are the ways of the intellect. We are not familiar with the ways of the intuitive, of surrendering to our higher selves, or relinquishing control so that we may begin to hear that still, small voice within.

Inner peace allows us to hear our inner voice. The inner voice is the voice of the higher Self, which will never tell us wrong. It is that part of us that is in intimate contact with the higher Self of all beings. It is our point of Oneness. When one possesses inner peace one knows the difference between the voice of the ego, which urges us to action based on security, sensual, and power needs and the inner voice of the higher Self, which seeks to restore the harmony and balance in our lives.

There are numerous avenues to the state of inner peace, including meditation, visualization, contemplation, dream recall, spiritual practice, Tai Chi Chuan, certain healing practices, massage, bodywork, etc. Whatever your path, it is one of the most important journeys you will ever take. It is perhaps, the *only* true journey.

What is Healing?

(Reprinted from *Crane Mountain Abbey Quarterly*, Fall 1980)

In seven years of involvement with healing, of working to heal myself and encouraging and facilitating others to do the same, I have had several different views of what healing is.

I think my initial understanding of healing, influenced heavily by our western medical model, was that it involved the elimination of symptoms—making people feel better. Although my belief in what works had shifted from prescription drugs, patent medicines and surgery exclusively to include good nutrition, herbs, and natural healing techniques, the goal was still the same: to make people feel better.

A summer spent as a sort of apprentice to Rolling Thunder, a native American medicine person in Carlin, Nevada, dropped the bottom out of my naive and superficial view of what healing is all about. The most important lesson I learned while I was there was that physical illness stems from some out-of-balance condition of body, mind, and spirit. When someone violates her own belief about what is right, whether it be a matter of what she believes to be a healthy diet, or her beliefs about what is ethical behavior, disease is a likely result. This may manifest in physical, mental, or emotional symptoms, and before a complete healing is possible, balance must be restored. If I violate my own beliefs about proper diet, I must either return to eating "properly," or take a look at my beliefs about food. In this age of many conflicting nutritional systems, it is easy for us to hold conflicting beliefs simultaneously (an example: yogurt is good for me because it maintains a healthy balance of digestive enzymes and provides protein, AND yogurt is bad for me because dairy products cause mucus). But native American medicine places a far greater importance on harmony in the area of right conduct. Rolling Thunder told me the story of a young man who came to him, very ill, offering to pay anything to be healed. The healer told the man, a recently returned veteran of Viet Nam, to go back to that country and make restitution to the people there for the death and destruction he had caused, and *then* come back for healing. His feeling about this man was that, deep within him, he was suffering from guilt for his actions in Viet Nam, and until he had laid this to rest, he would not be well.

After having observed and experienced Rolling Thunder's approach to healing, my understanding of what healing is had broadened and deepened; my definition of healing became a state of harmony of body, mind, and spirit, and, further, harmony with other beings and the earth. This remains a very basic and important part of my perception of what healing is.

Another dimension of healing has slowly unfolded for me which is directly linked to the growth of my spiritual awareness. What I have observed, both in myself and in people with whom I have worked, is that our ability to heal ourselves is directly related to our ability to love ourselves. It is as though self-hatred and self-condemnation block out healing energies.

The more I work with healing energies, the more I see them as the same as love, not in the sense of personal attachment, but rather the respect and appreciation of each individual manifestation of the divine. And if healing energy is indeed love, it seems that self-hatred would neutralize or nullify any such energy entering the body. (I think there are exceptions to this: magical, mystical moments when all such negative feelings are swept away by the power of love. But can the effects of this power endure if attitudes toward the self are not transformed?)

There is another aspect of love as healing which I think would have surprised me if I had considered it several years ago. It is the healing that comes with loving and giving of ourselves to others. I was raised with models that equated self-sacrifice with love; then, in early adulthood, I rejected those models because I saw them as martyrdom and manipulation of others. I was suspicious of "unselfishness." But I didn't understand the reality of joyous service, of giving for the love of it, and for the joy of seeing others bloom upon being loved and ministered to.

I worked with a woman for three years whose life was a powerful demonstration of the healing power of loving self and others. Michelle had been diagnosed as having multiple sclerosis, and when I first began working with her (doing massage and counselling), she was leading a life of quiet desperation. She was surrounded by negativity, rarely leaving her dark, gloomy house. She had great difficulty showing the love she felt for her young son, and almost all her attempts at communication with others came out as criticism and irritation. Her body was weak and filled with pain.

As she began to explore questions of living and dying, and to make choices about what she wanted for herself, she became more interested in healing and the possibilities for her own condition.

She discovered the power of believing she could create her own reality. As she began to think more and more positively about herself and her life, she related more positively to others; her pain also went away, and she became stronger. She moved away from her dark house and into the sunshine, and soon began to involve herself in working with others with M.S. She was a prime mover in a rap group for people with M.S., served on the board of a center for the handicapped, and gave free pottery lessons to handicapped people. The more she gave, the more she was capable of giving. She stayed carefully tuned to her body and her own inner voice to make sure she didn't overstep the bounds of her own well-being.

Michelle transformed her own being, and her healing process continued; her connectedness to others and to the Source of love and healing became a part of her everyday life. She was an inspiration to me, and contributed greatly to my present understanding of what is healing. If I were asked to sum up that understanding in a very few words, they would be these: love is healing.

Lila Forest

What You Can Do: Bringing Inner Peace to Your Life

There are many, many tools available in our culture that are useful in moving toward inner peace, because so many are seeking it. We give you a generous helping of suggestions here to stimulate your own creative flow and perhaps to introduce you to some new ideas. Many thanks to Nancy Binzen, Mark Thurston, Irv Thomas, Bill Cane, Margaret Boarman, Rio Olesky, and Jeremy Taylor for valuable contributions to this section.

Attitudes

Tell the truth.

Be honest with yourself and everyone.

Do what you say you'll do. (This has two parts: don't take on more than you can handle, and keep the commitments you make.)

Let other people be who they wish to be, and enjoy them.

Love and accept yourself just as you are.

Be willing to know your fears (death, pain, loss, etc.).

Realize that guilt is useless to you and resolve to catch it at work and to banish it from your life.

Discipline

Meditate (i.e., become still and listen) daily.

Remember your dreams and pay attention to their gifts to you (take a dream class to get you started).

Do the *Course in Miracles*.

Pray daily (i.e., talk to God).

Pray for people who awaken disturbing emotions when you think of them.

Spend time alone with nature.

Listen to your inner voice and follow it. (Try getting relaxed, close your eyes, say you want to hear your inner voice, and then listen. Do what it tells you/you tell you.)

Periodically notice your breathing and any bodily tension; take the time to tune into what's disturbing you and breathe into the tense places—i.e., inhale and then send your out-breath to the tension to relax it.

Do some soul-searching: are your goals and ideals your own, or what someone else thinks (or thought) they should be?

Do hatha yoga; make friends with your body. (Find a teacher, or if there are none around, read *Yoga, Youth, and Reincarnation*—see bibliography.)

Find one or more art media that you enjoy; let your inner self speak through color and form.

Ground in the light (i.e., realize your true nature as a spiritual being, and derive your strength and security from Spirit).

Do t'ai chi chuan; find your graceful, flowing self.

Receive a massage regularly.

Learn to visualize and create positive pictures within your being to sustain and inspire you.

Write affirmations and say them three times before going to bed. They can be very specific, such as, "I am free of smoking," "My self-confidence is increasing every day," or more general and universal, such as, "I am a totally healthy, loving being," or "I am one with all." (Avoid the use of negatives such as not or non- or un-; keep your affirmations as simple as possible. Getting clear about what you want is an illuminating process!)

Chant alone or with others frequently. Try Om, or whatever names of God are in your heart, or your own name.

Learn self-hypnosis from tapes.

See a spiritual counselor.

Read books. Try some of those listed in this book's bibliographies. Also try going to the metaphysical, spiritual, inspirational, or psychological section of a bookstore and let a book find you.

Keep a personal journal. Use it to work through difficult places, to record your joys, to ask yourself questions and then answer them, to dialogue between parts of yourself or with God. Consider writing your dreams in your journal (perhaps in a different-colored ink)—the connections between the two are important and enlightening.

Use the Earthstewards Network tapes as teachers and guides for your spiritual journey.

Let these suggestions serve as a stimulus to your own creative ways of moving toward inner peace; share your ideas with others, too. May your journey be blessed.

Bibliography

Section I (Inner Peace)

Thurston & Puryear, *Mediation and the Mind of Man,* ARE Press
Szekely, E. B., *The Gospel of the Essenes,* C. W. Daniel Co., London
dePurucker, G., *Golden Precepts,* Theosophical Publishing House
Parry, Danaan, *The Essene Book of Days,* Sunstone Publications
Philips, Howes, Nixon, eds., *The Choice Is Always Ours*
Faraday, Ann, *Dream Power*
The Dream Game
Jung, Carl, *Memories, Dreams, Reflections*
All the Edgar Cayce material (see ARE in Appendix for catalogue)
The Bible (pick your version)
The Bhagavad Gita (any translation)
The I Ching (try the various versions)
The Tao Te Ching (many translations)
Ponder, Catherine, *The Dynamic Laws of Healing*
Thoreau, Henry, *Walden*
Prather, Hugh, *Notes to Myself*
Gibran, Kahlil, *The Prophet*
Merton, Thomas, *The Seven Storey Mountain,* many other works
Krishnamurti, J., *Think on These Things*
Awakening of Intelligence
Teresa of Avila, *Autobiography*
St. John of the Cross, *Dark Night of the Soul*
Meher Baba, *Discourses:* Vol. 1-4
Hesse, Herman, *Journey to the East*
Siddartha
The Glass Bead Game
Reps, Paul, *Zen Flesh, Zen Bones*
Lindbergh, Anne M., *A Gift from the Sea*
A Course in Miracles
Stearn, Jess, *Yoga, Youth, and Reincarnation*
Progoff, Ira (works on the Intensive Journal process)
Taylor, Jeremy, *Dreamwork*
Khan, Hazrat Inayat, *The Sufi Message* (13 volumes)

Section II:

Earth as a living organism

When we recognize that our planet itself is a living organism coevolving with humankind we shall become worthy of stewardship

Gaia

In what was perhaps the most visible single act of expanded consciousness in our time, an astronaut was able to stand back from our planetary home and view it in its entirety, its wholeness. He not only saw national and geographic boundaries disappear; more importantly he experienced the oneness of focus of all beings on that planet. Our earth was experienced as much more than a sphere of dead rock populated by various competing life forms; it was one body, one entity, one conscious being made up of billions of parts. Some of those interrelated parts were themselves conscious of their unique existense within the whole, some were not.

This breathing, pulsating totality, this living organism earth, is the subject of a refreshingly new way for the human species to view itself within the perspective of the whole. In his book, *Gaia: A New Look at Life on Earth,* J. E. Lovelock describes our planet as a living being, co-evolving with humankind. Just as we are climbing the evolutionary ladder in terms of our physiology and our consciousness of who we are, so too is Gaia (the ancient Greek word for the Mother Earth Goddess). Earth is, through the evolution of human awareness, involved in its own evolution of awareness, from the beginnings of simple, self-regulatory functions to the point of self-consciousness. In other words, the planet is evolving towards the day when it will be aware of itself, and our human awareness of self and of oneness is a major part of that process.

The Gaia hypothesis sees every life form on earth as an important "cellular" part of one self-regulating, living organism. And that organism is changing, growing, becoming more and more conscious. The concept raises such questions as these:

Is our collective intelligence the "brain" of Gaia?
What is the proper function of humankind within the evolving consciousness of the larger organism?
When a species is removed from the earth, what is the effect on the whole?

The Gaia hypothesis may be an important advance in environmental thinking; it lifts our limited perspective to the clear vantage point of Whole Earth and makes the awareness of our Oneness a global imperative.

Living the Rhythm of Nature

(Reprinted from *Spirals*, No. 6, August 1978)

The Biblical expression, "to each thing there is a season," has been popularized in song, but to what extent is its wisdom practiced? In today's world of frozen food and pep pills, we seem to be constantly competing with the subtle yet powerful cyclic rhythms of nature that call to us.

The ancient peoples recognized not only their connection with nature but also the effect that the rhythms of nature had on their bodies and their lives. They did not ignore or compete with the cycles of the seasons and the forces of nature. They *were* nature. Is it not so for us as well?

Using the Taoist metaphor, we may see ourselves as cherry trees. Sprouting, growing, bearing fruit, withering and finally returning to the earth to nourish it so that other life may grow.

More immediate than this, our "cherry trees" experience the cold of winter on our bare branches, the budding blossoms each spring, the rich fruit of each summer, yielding to the windy tug of autumn on our browned leaves. All this I can feel in my body if I quiet down and re-tune with the flow of nature.

I live in a community that endeavors to gear our lives to this flow of nature, and the experience is a rich one.

Autumn for us is a time of preparation, of slowing down and gathering together for support. At our Autumn Equinox celebration we share the fruits and nuts of harvest, and the chanting lasts into the night.

Winter is a season of introspection, of contemplation, a time for "mending nets." At Midwinter Solstice we light the Yule for warmth and we re-tell legends. The seed grows deep in the earth.

Spring. Initiation. A time for the seed to reach towards the new sun and to blossom. Our Vernal Equinox ritual sees some of us skinny-dipping in the cold Pacific. Contact with one another and with the earth, movement from within to without.

And Summer. The time for maximum outward creativity. Building, milling, digging. At Midsummer Solstice, the cornucopia overflows. Dancing, music, the delight of children of all ages.

The point is simply this: our bodies are a part of nature and respond to the differing energies that are present at the changing seasons. We can choose to align our lives with the flow of nature, with the Tao, and experience the subtle forces of nature within us, or we can compete with this rhythm. There are a great many ailments in our society today that I could connect with this course of "competition with the rhythm of nature."

I prefer to accentuate the positive and to invite you to discover for yourself the harmony of your seasonal flow. What are *your* asociations with the different seasons? Are you working like crazy in winter and lying about in summer, when your body is telling you that it wants to enter a meditative state in winter and be physically active in summer?

We have only to listen. Inside, there lie all of the answers to our questions, if we can be still enough to hear.

Danaan Parry

What You Can Do: Celebrating the Seasons

One of the very best ways to begin to tune in to the earth and her cycles is to mark the high points of those cycles with some kind of celebration. A celebration can be as simple as one person singing a song to the full moon or taking a walk in its light, or as elaborate as a large group of people performing a well-planned ritual followed by feasting and merriment. The purpose of this article is to get you started on the road to celebration and to encourage you to use your creative energies to link yourself more consciously with the natural order.

Some General Thoughts About Celebration

The most important thing to keep in mind about celebration of the seasons is that its purpose is to connect you to the natural world in which you live. Therefore, there should be a very personal element in every celebration, giving each person a chance to reflect on, and possibly share, how the particular quality of the event being celebrated affects and is reflected in her/his being. In an individual ceremony or one involving a small group, this is easily done. For a group of 30-100 people, a toasting cup or other symbol of one's turn to speak may be passed around a circle, giving each person an opportunity to contribute personally (and briefly) to the ceremony. For larger groups, the sharing may be done symbolically, such as writing it on a slip of paper or investing it in some small object, to be burned, or planted, or whatever might be appropriate to the occasion.

Music is very important to ritual and celebration, whether it be singing, chanting, live instrumental music, or rhythm-making.

It is better to sing a simple song that everyone can learn than to do a complex one that no one feels sure of. Music can be the carrier of spirit if it comes from the heart, and this can only happen when we don't have to think about what comes next.

In planning a ritual or ceremony, the following elements might be considered:

1

Salute to the four directions—here are some aspects associated with each:

EAST—sunrise, wisdom, dawning of the light
SOUTH—warmth, fertility, love
WEST—sunset, reflection
NORTH—cold, illumination, purification

2

Blessing with the four elements (sometimes done by walking around the circle with them)—here given with symbols for them and related aspects of the human being:

EARTH—salt—body
WATER—water—emotions
AIR—incense—mind
FIRE—candle—spirit

3

Some kind of personal involvement, as discussed above.

4

Music and/or dance

5

Ritual of sharing of food or drink

If You Have Never Planned a Celebration

About two months before the event:

1

Get together with a few other people who are interested. Brainstorm on the meanings or qualities of the event; list them on paper. This will give you a record of the feelings you want to evoke or what you want to celebrate. When this is clear, you can begin to give form to the ceremony and activities you want to have as part of the celebration.

2

Using suggestions in this article and any other materials you have access to, plan the elements of the day and the ceremony. It is important to have a balance of organization and room for spontaneity, both in the ceremony and in the activities for the whole day. Every gathering has its own unique spirit, and if there are opportunities for that spirit to be expressed, your event will have meaning and power.

3

Choose a location. This can be on private property to which you have access, or in a park. Find out if you need to reserve a spot or get any kind of permit. Decide if you want the event indoors or outdoors. The energy will be very different in one than the other. Outdoors gives more sense of connection with nature; indoors focuses energy more in toward the center and fosters connection with each other and with the intangible.

4

It's best to have your first celebration advertised by word of mouth and/or through an organizational newsletter, rather than by invitation to the general public. If your celebration is meaningful to those who attend you'll be amazed at the number of inquiries about the next one! You want to attract people who want to renew their connection to the natural and the spiritual, rather than those who want only an excuse to party. Don't worry about this—if the purpose and vision of you, the creators of the event, are clear, you will attract the people who will manifest and enhance that vision.

5

Spread out the responsibility for the various aspects of the celebration among as many reliable people as necessary, so you avoid overburdening one or a few organizers. Everyone should enjoy the event, especially those who have given it birth! Some elements that might need a facilitator are parking, food (everyone who comes should contribute), ceremony, and other activities such as dancing, singing, games, etc.

6

If there were expenses incurred in bringing about your event, don't be afraid to pass the hat! People are more than willing to contribute to a well-organized, meaningful, and enjoyable happening.

7

Get people to clean up after themselves, and nab a volunteer cleanup crew to finish up, before everyone disappears.

Some Times to Celebrate

There are a number of occasions that one might choose as a good time to hold a celebration; here are the ones that we observe.

The solstices and equinoxes are primarily solar events, celebrating different phases of our experience of the sun. As such, they have a masculine or yang quality and have traditionally been associated with male gods. (Paradoxically, while this is true, it is also true that winter solstice is the time of greatest yin and least yang, while at summer solstice the polarity is reversed.) The cross-quarters (Imbolc, Beltane, Teltane, Samhain) are focused on the earth and her cycle and thus partake of the feminine or yin aspect. The new and full moons are naturally lunar events and also yin in nature.

Under the name of each of the eight points in the yearly cycle (solstices, equinoxes, and cross-quarters) you will find a list of holidays observed in our culture which bear direct relationship to that point in the wheel of the year and which occur close to it. This will help you to see how the deep currents of celebration have flowed into our era, even though the forms have changed and the apparent reasons for celebrating are different.

Also given for each of the eight points is a seasonal focus, indicating a stage in the cycle of plant life. This cycle is also symbolic of an inner, psycho-spiritual cycle that is part of our experience as human beings on the earth. We would do well to heed this cycle and to take its phases into consideration as we make choices about how to be and what to do throughout the year.

For each of the eight events, suggestions are given for ceremonies, colors, plant materials, incense, and food and drink to be used in celebrating them.

Seasonal Celebrations

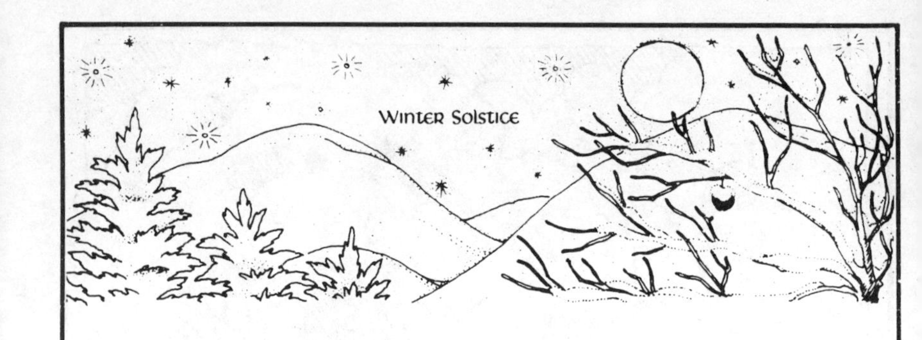

Winter Solstice

(Christmas, Hanukkah)

The seed stirs in the earth

The essence of celebrating Winter Solstice (sometime between December 19 and 23, when the sun enters Capricorn) is the honoring of the darkness, which is at its peak, and joyful acknowledgment of the return of the sun (the Son). Many of the traditional customs once observed at this time throughout the western world have been incorporated into the Christmas observance, such as the "Christmas" tree, mistletoe, holly, use of red and green, and the yule log.

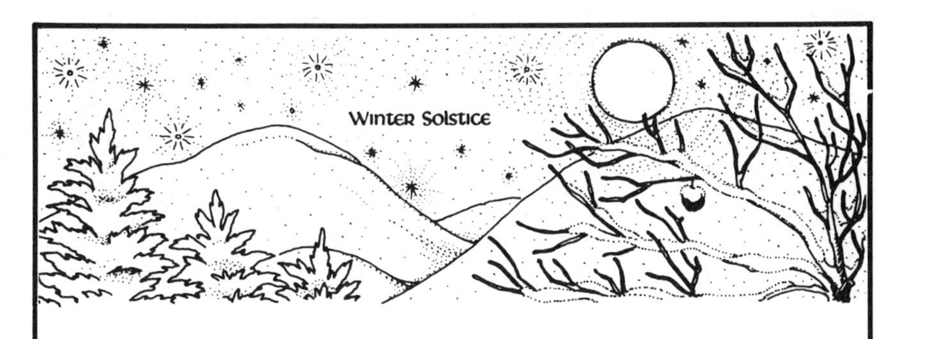

Some suggestions:

Extinguish all lights and fires in your house before the longest night. Light a fire in the morning and place on it a decorated Yule log. Later, give a piece of it to everyone present with which to start the first fire of the reborn light at home.

Hold a vigil through the longest night, chanting, singing, meditating, praying, taking turns reading.

Light a candle each night from Solstice to Christmas—feel the connection between the rebirth of the sun and the birth of the Son.

COLORS: Red, green, white, black
PLANTS: Holly, mistletoe, evergreens, berries
INCENSE: Frankincense, myrrh, pine, frangipani, bayberry
FOOD AND DRINK: Blazing puddings, wassail, mulled wine, fruitcake, cranberries

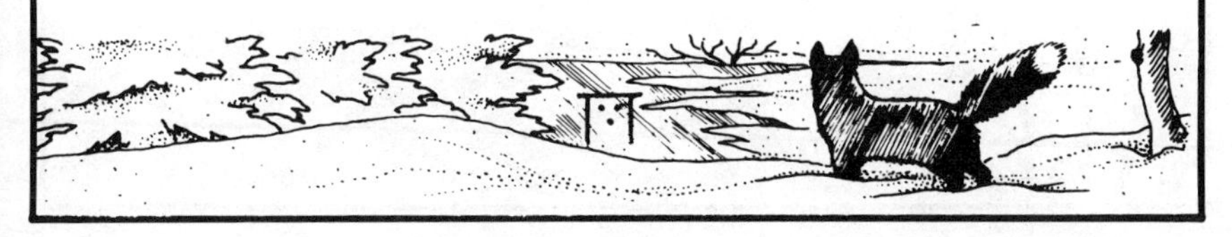

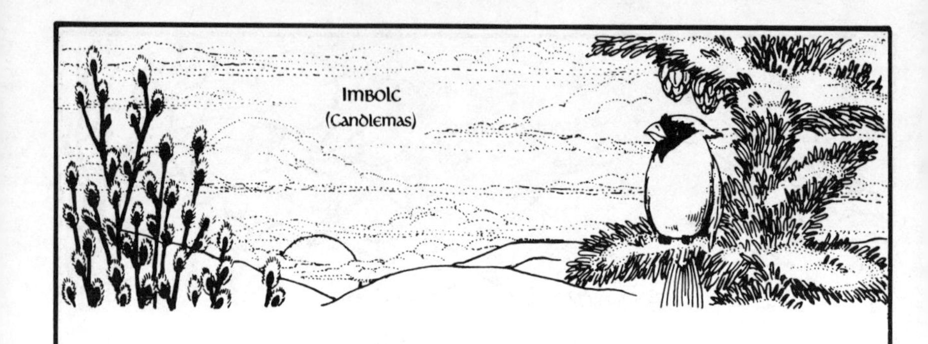

Imbolc (February 1)

(Lady Day, Candlemas, Groundhog Day, Valentine's Day)

The new leaves unfold and buds appear

Imbolc is approximately halfway between Winter Solstice and Spring Equinox. In Celtic lore it is a fire festival, and the day of the washing of the earth's face. Christianity has picked it up as the day of the dedication of the Virgin Mary in the temple, and a festival of candles.

In ancient days it was the day of making and breaking commitments of marriage, a contract that lasted a year. In warm climates, it can be considered the first day of spring.

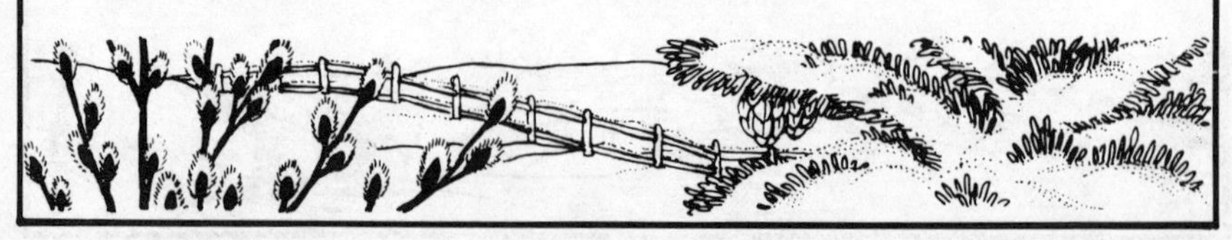

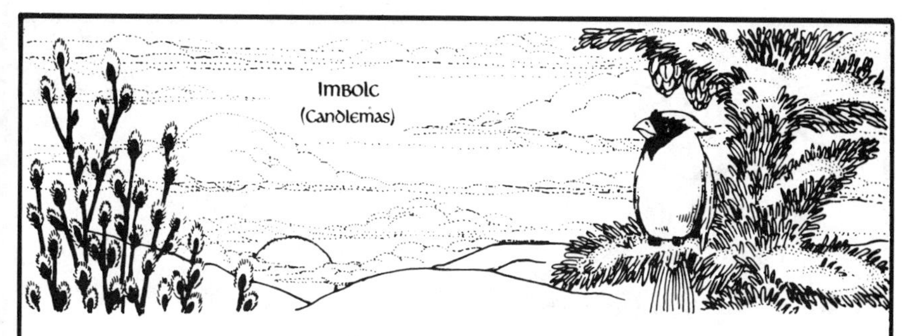

Some suggestions:

Build a big bonfire and sing songs to the coming spring.

Renew your commitment to your partner—invite other couples to join you.

Ceremonially acknowledge the ending of a partnership.

Bless candles to be used in ceremonies and for meditation in the coming year.

Reflect on the tender young shoots growing up within you—share them with friends.

Make a commitment to something for the next year.

Have a candlelight procession in the park.

COLORS: white, grey, pale green, pastels
PLANTS: grass, early bulbs (daffodils, crocuses, jonquils)
INCENSE: juniper, cedar, eucalyptus, pine
FOOD AND DRINK: sprouts, oatcakes, custard

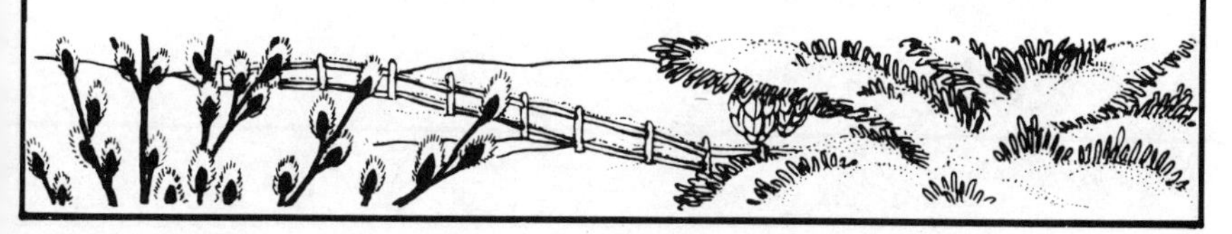

Spring Equinox

(Passover, Easter)

Flowers announce the promise of fruit

The day when the sun enters Aries (which varies from March 19 to 23) is Spring Equinox. Days and nights are of equal length, and things are blooming. Traditional Easter customs predate the coming of Jesus: flowers and eggs representing fertility, dressing up in new clothes. Interestingly, the date of Easter each year is set by determining the first Sunday after the first full moon after equinox. The word "Easter" derives from Oestre, a fertility goddess. All is one!

Some suggestions:

Choose a budding girl/woman to be Spring—crown her and sing to her.

Have the children give flowers to everyone or shower them with petals.

Sing spring songs.

Dance line and circle dances.

Reflect on the colorful flowers you give the world.

Hold rites of Spring with observances meaningful to you beginning with Equinox, honoring the full moon, culminating in Easter or Passover. (If Passover precedes Equinox, do it the other way around!)

COLORS: all the colors of the flowers
PLANTS: blossoming branches, tulips, iris, all spring flowers
INCENSE: flowery fragrances
FOOD AND DRINK: decorated boiled eggs, light breads flavored with anise or cardamon, Rhine wine

Beltane (May 1)

(Mayday)

The fruit ripens

Beltane is about halfway between Spring Equinox and Summer Solstice. It is the celebration of full flower, of the dance of male and female. It is celebrated the world over with joy and merriment. In warm climates it is the first day of summer.

For thousands of years this day has been celebrated as the point of fertility, as a time when Nature blatantly displays its beauty to bring about the conception of new life. Beltane, or Bealten, means the magic of flowers in the Celtic tradition.

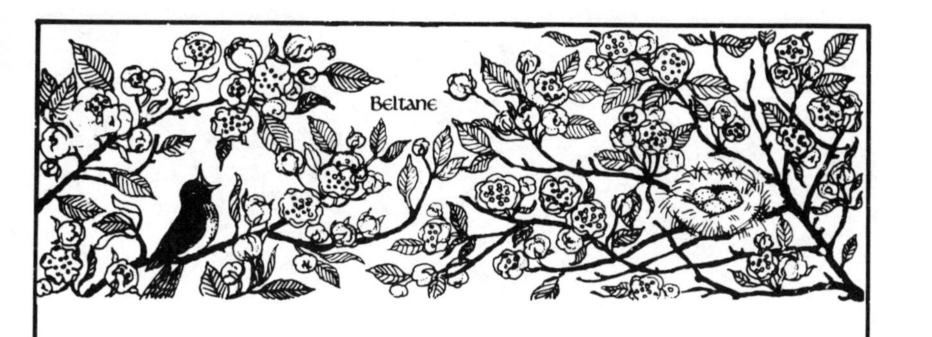

Some suggestions:

Divide your group into men and women—sing to each other.

Have a procession of flowers.

Put up a maypole with streamers and dance around it.

Reflect on how well the feminine and masculine energies within you are balanced.

Choose a Pan and a Queen of the May—put on a play, or play with the roles.

COLORS: all the bright colors of flowers
PLANTS: all flowers
INCENSE: jasmine, flowery scents
FOOD AND DRINK: May wine, strawberries, fruits from your area, oatcakes, custard

Summer Solstice

(Fourth of July)

The tree bends with ripened fruit

Summer Solstice, when the sun moves into Cancer (June 19 to 23), is the all-out celebration of the sun, of long days and warm evenings, of life out-of-doors with sun and water and blue skies.

This is when the sun exerts its maximum power upon our part of the earth as its rays strike us head-on. It is a time of full outward physical manifestation. The powers of inner contemplation are at their lowest point, and everywhere are the energies of "doing", of exerting the will.

Some suggestions:

Combine a ceremony to honor the sun with an old-fashioned picnic or barbecue, complete with swimming, roasting marshmallows, and singing around a campfire.

Reflect on the light you bring to the world and the outward results of your creative nature.

Bless the products of your vegetable garden.

Sing praises to the sun.

Paint each other's faces.

Make ice cream.

COLORS: yellow, orange, bright colors
PLANTS: flowers (especially sunbursts)
INCENSE: sandalwood, spicy scents
FOOD AND DRINK: fresh vegetables and fruits (especially home-grown), homemade ice cream, barbecue, homemade pies

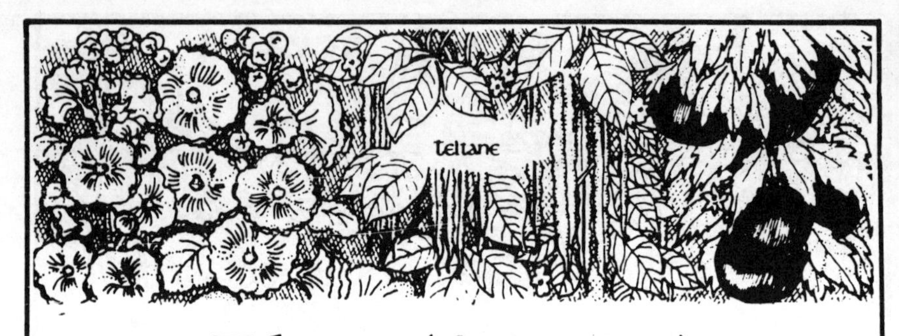

Teltane (August 1)

The bountiful harvest is gathered

Teltane, halfway between Summer Solstice and Autumn Equinox, is a Celtic fire festival. It was traditionally a time for games similar to the Olympics, and it was considered the first day of autumn. The purpose of the ceremony was to marry the sun to the earth and thus rejuvenate the sun until the harvest was complete. It is also a celebration and offering of the first fruits of the harvest.

On Teltane, or Lammas, or Lughnassad, Lugh, the Celtic God of Light, buried his foster mother Tailltiu beneath a great mound in Ireland. This signifies the withdrawal of the Mother Goddess into the earth in preparation for the falling seeds that will need her care if they are to germinate in winter.

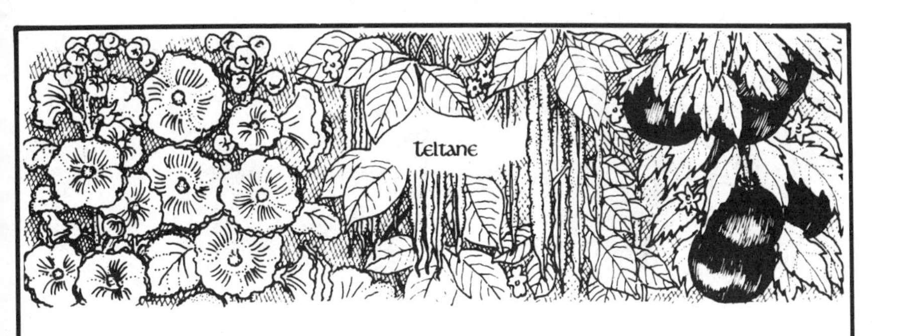

Some suggestions

Have a playday, with New Games, volleyball, folk dancing, races.

Have a marriage ceremony for the earth and sun.

Recognize the fruits of your growth and your labors.

Thank the earth mother for all her gifts.

COLORS: red, gold, bright colors
PLANTS: flowers
INCENSE: spicy scents
FOOD AND DRINK: same as Summer Solstice–oatcakes, custard, beer

Autumn Equinox

(Yom Kippur)

The harvest is stored

The sun enters Libra (September 19-23)—once again the days and nights are of equal length. This is a time to give thanks for the fruits of the earth and to prepare for the inward time that is coming. In the Jewish tradition, the new year begins at this time.

This is the point on the wheel of the year when there is balance between the energies of outward, physical, yang manifestation and inward, psychic, yin creativity. It is a time to surrender to our inner nature as we move toward winter.

Some suggestions:

Make a mandala of grains and seeds on the ground, an offering of earth's gifts to the earth and her children.

Reflect on your achievements and creations of the past season and on the quieter time ahead.

Honor the elders.

COLORS: gold, rust, scarlet, yellow, brown
PLANTS: chrysanthemums, sheaves of wheat, autumn leaves
INCENSE: spicy
FOOD AND DRINK: grains, seeds, nuts, apples, raisins, dried fruits, wine

Samhain* (November 1)

(Hallowe'en)

The tree stands bare and the seed lies still in the earth

Midway between Autumn Equinox and Winter Solstice, Samhain is the one night when the world of form and the world of spirit touch, thus all our Hallowe'en associations with ghosts and goblins. A time of long rays of the sun giving mellow light and gentle warmth to the waning days.

Traditionally, it is the first day of winter, a fire festival for burning away drosses in preparation for the rigors of that season.

Some suggestions:

Honor all your loved ones who are in spirit, your guides, the spirit of that which you aspire to.

Hold a seance.

Have a ritual fire in which you symbolically burn your fears and worries – purify yourself for winter.

Do some divination with the I Ching, tarot or your favorite oracle.

COLORS: gold, rust, scarlet, brown, gray
PLANTS: pumpkins, autumn leaves, sheaves of grain, grasses, dried plants
INCENSE: cinnamon, patchouli
FOOD AND DRINK: wine, squashes, pumpkin, nuts, seeds, apples, dried fruits, grains

**Pronounced SOW-WAIN*

The Moon Cycle

The moon's cycle through each twenty-eight days mirrors the birth, life, death and rebirth of all things. The two points in that cycle which are often marked by celebration are the new moon and full moon, the qualities of which are very different.

New Moon

The essence of this time is newness—all things are possible. Celebrate new beginnings, make new commitments, initiate new projects. Mark this day as the beginning of a two-week period of growth and increase, culminating in the full moon.

Full Moon

What a magical night! The full moon is a culmination, a fullness, a maximum stimulation of our intuitive, a-rational, creative, sometimes-a-little-crazy aspects. This is a wonderful time to connect psychically with people you love who are far away. The moon is full all over the world tonight—use it to send messages and thoughts of love. (Of course, nighttime is different from one side of the globe to the other, but when night comes, the moon will be full.) A circle of people under the full moon have the perfect opportunity to experiment with their psychic connection to each other, which can be maximized by silent, meditative attunement, or by chanting, singing, humming (or even howling) together. A time for seeing visions, telling dreams, sending healing energy to those who need it. A time to revel in and surrender to your yin or feminine aspect. (Each of us, male and female, has both a masculine and a feminine side—our task is to fully develop both and enable them to live in love and harmony within us.) Experiment with celebrating in groups of men and women together and in groups of the same sex. The energies are very different!

Some suggestions for elements of lunar celebration:

Candles, incense, shimmery colors (white, opal hues, silver), healing, singing, dancing, meditation, blessing each other, sharing visions. Try drinking mugwort tea (also known as artemisia, sacred to the goddess of the moon); it has the reputation of stimulating dreams and visions. Bake a round or crescent-shaped loaf of bread to share. If you're in a women's circle, celebrate your menstrual cycle, so linked to that of the moon. If you're a man, become aware of and celebrate your own emotional cycle, whatever it is. Find your own way of honoring the orb that influences our lives so powerfully from day to day.

Conclusion

Use all this information and the suggestions given as grist for the mill of your creative self—there are no shoulds in this game! Let your imagination, your feelings, your fantasies, play an important role in the process of creating a celebration.

We wish you well as you launch into this wonderful world of celebration. May it be as powerful, as enjoyable, as inspiring for you as it has been and is for us.

Bibliography

Section II (Earth as a Living Organism)

Lovelock, J. E., *Gaia: A New Look at Life on Earth*
Parry, Danaan, *Essene Book of Days,* Sunstone Publications
Goodman, Jeffrey, *We are the Earthquake Generation*
Graves,Tom, *Needles of Stone*
Michell, John, *The Earth Spirit*
Secrets of the Stones
Watson, Lyall, *Supernature*
Spangler, David, *Festivals for a New Age*
Niehardt, J. G., *Black Elk Speaks*
Szekely, E. B., *The Gospel of the Essenes*
Heline, Corinne, *Mystery of the Christos,* New Age Press
Myers, Dr. Norman, *Gaia*
Cohen, Alan, *The Healing of the Planet Earth*
Seielstad, George, *Cosmic Ecology*
In Context magazine, the "Art and Ceremony" Issue 4

Section III:

planetary stewardship

"We looked down and saw planet Earth and realized that it was our beautiful and very small home. We realized things were not going so well down there and we felt we had to do something about it."

Astronaut Edgar Mitchell

When we see ourselves as stewards of our planet & not as owners & masters of it there shall·be lasting satisfaction from our labors.

Planetary Stewardship

From ancient oneness

Through a time of separation,

And now, an evolution to the point

of conscious oneness —

This appears to be our path . . .

The skies are filled with great white geese migrating homeward after winter's southern sojourn. Soon they will return to their ancestral mating marshes high in the Canadian wilderness. Are these wondrous beings thinking about which bearing to fly, or worrying about being off course? It is I, watching them, who hold such thoughts, not they. For they possess an instinctual knowing that I have long ago surrendered so as to develop the faculty with which I ponder their being "on" or "off" course. They are at one with the world in which they fly, and they do not wonder why.

Early in the development of humankind, we too felt our oneness with all that is, as we naturally and unconsciously lived each moment in harmony and balance. As do the few remaining aboriginal humans today, we wandered in the spirit, being given that which was needed, and feeling our symbiotic, psychic connection with every living thing. The concept of "reverence for nature" would have been ridiculous, since, we *were* nature, not apart from it. Although a unique species, we still retained the connection to all that is.

As each kind of living thing evolved so as to express its unique gift within the whole, humankind began to develop its gift, that of being conscious of its-SELF. For us, the task was to evolve to the point where we could integrate the paradox of being: at the same moment separate and yet connected, unique and yet one.

And so, through thousands of years of evolution, we have journeyed far from our original state of *being*, from our surrender to our oneness with all. We have foregone our intuitive knowing nature, while our physical bodies and our minds have developed their uniqueness, their separateness, so as to fully experience all that there is to experience. And while this has been quite perfect in terms of the necessary development of the "separateness" pole of the paradox, still we must remember that our task as conscious beings is not only to *experience* uniqueness but also to stay conscious of our Oneness with the All.

As we look at the state of our physical world today, we see the effect of our lack of remembering. When we forget that we must manifest our total connectedness with all beings at the spirit and heart levels while we manifest our uniqueness at the physical level, then an imbalance occurs. And because of our ability to create our reality, this imbalance can bring us to a point of ultimate separation of poles, to a point of destruction of our unique life forms.

This lower form of return to oneness is not in harmony with the unfolding plan of which we are a part. The higher path, which is available to us if we choose to accept it, is a return to oneness through conscious evolution. That is, we have within our range of alternatives, that of evolving to a point of Conscious Oneness.

This is the completion of the circle, except that it is now a spiral.

To choose the spiral path of upward conscious evolution rather than the circular path of locked-in movement from oneness-to separateness-to devastation-to oneness, all in an unconscious manner, we must accept the challenge that we ourselves have set before us. That challenge is to remember. We must remember who we really are, in the midst of acting out who we arranged to be. This is the paradox of oneness and separateness.

Each of the unique manifestations of the One has a role in this unfolding drama of spiraling evolution. Our role, no more or less important than any other, is to develop, and use our consciousness-of-self as a tool for service. Humankind is unique among all beings primarily in that we are conscious of our consciousness.

From this awareness has come the concept of Planetary Stewardship, as a vehicle for the full use of our unique gift. One definition of Steward is ". . . one of the servants on a passenger ship." Planetary Stewardship may be defined as those servants on spaceship earth who have accepted the responsibility of employing their expanded consciousness in the service of the whole. Stewards own no-thing, yet they care for all things. Planetary Stewards are those humans who are beginning to remember that we are here not only to experience our uniqueness, but also to utilize that uniqueness in the service of Oneness. From oneness, through a time of separation, and now an evolution to the point of *conscious* oneness — this is the path of stewardship. When we see ourselves as stewards, carers, lovers of our planet, and not as owners and masters of it, then shall there be lasting satisfaction from our labors.

The White People Never Cared for Land

(Reprinted from *Religious Perspectives in College Teaching*)

The White people never cared for land or deer or bear. When we Indians kill meat, we eat it all up. When we dig roots, we make little holes. When we build houses, we make little holes. When we burn grass for grasshoppers, we don't ruin things. We shake down acorns and pinenuts. We don't chop down the trees. We only use dead wood. But the White people plow up the ground, pull down the trees, kill everything. The trees say, "Don't. I am sore. Don't hurt me." But they chop it down and cut it up. The spirit of the land hates them. They blast out trees and stir it up to its depths. They saw up the trees. That hurts them. The Indians never hurt anything, but the White people destroy all. They blast rocks and scatter them on the ground. The rock says, "Don't. You are hurting me." But the White people pay no attention. When the Indians use rocks, they take little round ones for cooking . . . . How can the spirit of the earth like the White man? . . . Everywhere the White man has touched it, it is sore.

Wintu holy woman

What You Can Do: Applying Principles of Planetary Stewardship in Daily Life

As participants in twentieth century western culture, we must of necessity avail ourselves of some products and services that operate in violation of the principles of planetary stewardship, if we are to survive. We can work toward changing some of these, we can reduce our levels of consumption, we can do the best we can. It is important that we start from where we are and take whatever steps we can toward greater care of our ecosystem, and that we not feel guilty for the rest. Having said that, there are many small things we can do to move ourselves toward being more a part of the solution and less a part of the problem.

Find out where the nearest recycling center is, and recycle your used bottles, cans, and newspapers. This reduces solid waste disposal problems and slows down the use of non-renewable and renewable resources (metal ones, energy, and trees).

Become aware of excessive packaging, particularly of food, and of your disposables. Make minimal packaging part of your criteria for selecting purchases. Use cloth napkins, dishtowels, and sponges instead of paper towels and napkins; collect containers with lids for storing leftovers and minimize your use of plastic wrap.

Buy no- or low-phosphate detergents and cleaners. Wash dishes by hand (dishwashers use a lot of electricity and require high-phosphate detergents) or only wash full loads in the dishwasher.

Become conscious of what you dump into our planet's bloodstream (the waterways) through your sinks, shower, bathtub, and toilet. Move toward the use of natural, biodegradable substances.

Break the running water habit—while you wash dishes, shampoo your hair, brush your teeth. (Rinse dishes in running water for sanitary reasons.)

Instead of throwing out your food scraps, compost them and use the rich result on your garden, or offer it to someone who has a garden, or take it to an empty lot and restore the soil there. The simplest way to compost is to develop a heap in the corner of your yard, covered by a sheet of plastic during the rainy season — it pretty much does its own thing! For more detailed instructions and suggestions for apartment dwellers with no back yard, consult *The Integral Urban House* (see bibliography). There are also many other good books available in libraries and bookstores on organic gardening and composting.

Have an insulation consultant come from your local oil, gas or electric service and check out your house. Many offer this service free, advising you on ways to conserve heat in cold weather and keep cool in summer. The recommendations are detailed, practical and often cost very little to implement (example: an insulating blanket for your hot water heater costs about $10-$15 and pays for itself in a few months). Your house will be more comfortable, too.

Look in the phone book for these utility companies and call for energy conservation and solar information. If they don't provide this service, write them a letter and suggest that they do so; have your friends write, too.

Become aware of the danger of extinction posed to whales and seals by systematic hunting on a commercial scale; support organizations such as Save the Whales and especially Greenpeace, which works tirelessly and courageously to stop the slaughter.

If you have children, enroll them in the Ranger Rick Nature Club and receive *Nature Magazine*, a wonderful children's publication produced by the National Wildlife Federation (see bibliography). The articles and colored photographs of animals of all kinds are educational and inspiring, and they foster a sense of respect and responsibility for all the life forms on and in the earth.

Become aware of how energy is generated in your area, and keep informed about changes. Let the utility company know your opinions in hearings and letters.

Write in your journal about you as a caretaker of the earth and read what you have to say.

Plant a tree for every year of your life.

Plant a garden—vegetables and flowers.

Learn about Voluntary Simplicity (see bibliography).

Learn about methods for use of solar energy.

Become conscious about your use of fuel for transportation; find ways to decrease your consumption (bicycles, carpools, public transportation, advance planning to minimize errand-running and shopping trips).

Support groups working for de-militarization and alternatives to nuclear energy (see list of organizations at end).

Many of these suggestions are actually most effective in the way they affect your own consciousness when they are applied. The more you become aware of what your life support systems are, how they work, what the cost is to the earth, and what, if any, repayment they make to the environment, the more you are connected to the earth and, therefore, concerned about it. A basic attitude of concern is probably more important in the long run than some of the measures that seem difficult to apply. If every human being on earth had a basic concern for the well-being of the planet and all her life forms, the world as we know it would be transformed. So we start with ourselves and our children, increasing the scope of our awareness and doing what we know to do.

And for the rest? Try making a daily offering by visualizing the whole earth surrounded with the light of love, peace, and healing, touching all beings. And may it transform us all.

Bibliography

Section III (Planetary Stewardship)

Planetary Citizens, *Planet Earth* (journal,) see Appendix for address
New Alchemy Institute, *Journal of the New Alchemists* (see Appendix)
Farallones Institute Staff, Olkowski, Olkowski, & Javits,
The Integral Urban House
Elgin, Duane, *Voluntary Simplicity*
Brand, Steward, *The Next Whole Earth Catalogue*
Halifax, Joan, *Shamanic Voices*
Bateson, Gregory, *Steps towards an Ecology of Mind*
Mind and Nature
Castenada, Carlos, All of his books
Caldicott, Helen, *Nuclear Madness*
Hawken, Paul, *The Magic of Findhorn*
Pennick, Nigel, *The Ancient Science of Geomancy*
Lindisfarne, *Earth's Answer*
Publications from Friends of the Earth and the Sierra Club (see Appendix)
The *Foxfire* Series
McLuhan, T. C., *Touch the Earth*
Olkowski, Helga, *Backyard Composting*
Jeavons, John, *How to Grow More Vegetables*
Shuwall, Melissa, ed., *How to Shrink Your Energy Bills*
deBell, Garrett, ed., *The New Environmental Handbook*
Johnson, Warren, *Muddling Toward Frugality*, Shambhala, Boulder, CO
Nyerges, Christopher, *Urban Wilderness: A Guidebook to*
Resourceful City Living, Peace Press
Griffin, Susan, *Woman and Nature*, Harper & Row
Lee, Dorothy, *Freedom and Culture*
McLaughlin and Davidson, *Builders of the Dawn*
Theobald, Robert, *Rapids of Change*
Ram Dass and Gorman, *How Can I Help?*
In Context magazine, "The New Story" Issue 12
National Wildlife Federation, *Ranger Rick's Nature Magazine*, 1412 16th St.
NW, Washington, DC 20036

Section IV:

Right Livelihood

"Work is love made visible. And if you cannot work with love but only with distaste, it is better that you should leave your work and sit at the gate of the temple and take alms of those who work with joy. For if you bake bread with indifference, you bake a bitter bread that feeds but half our hunger."

Kahlil Gibran
(Reprinted from The Prophet*)*

Right Livelihood

In 1978, I was hired to conduct a values clarification workshop for 400 nurses in a major city hospital. To get started I asked them to write down the twenty most enjoyable ways that they spend their time. Of the 400 nurses, not *one* of them wrote in any of their twenty items *anything* that related to their work. Every one of them saw nursing as a means to an end, a job which gave them money so they could then enjoy themselves. Yes, there were many who viewed their work as a vehicle for service, but not for enjoyment.

Twentieth century philosophy has a basic tenet that work and play don't mix, and more subtly implies that work and self-fulfillment don't mix. Essentially, we "rent" our bodies and perhaps our brains to the highest bidder. In return, the lessor of these commodities provides us with money, a magical medium which we can use to buy enjoyment, relaxation, respect, survival, fulfillment. We tolerate the work because it provides the magical medium, and we most likely resent the lessor, or the system, that maintains this duality. But of course we are an intimate part of this system, a co-oppressor as well as the oppressed. This is so because of our attachment. Yes, it is an attachment that is supported by the entire structure of our society, and therefore difficult to alter. But we are dreamers, you and I, and we know that the moment is swiftly approaching where the structure will no longer be supported upon its decayed foundations. What then will we have to present as the plan for the new experiment? And how can we, in our lives right now, plant the first seeds of wholistic living? Perhaps one answer lies in the way we approach our livelihood.

It has been suggested that there are several steps in a person's path to right livelihood as different kinds of needs are met:

Level	Rewards	Needs met
JOB	Any job is okay; pay is what counts; work to live.	Survival (physiological needs: food, shelter, clothing)
PERMANENT JOB	Regular work at routine tasks; work to stay secure, vacations and retirement important.	Security (satisfy basic needs over time)
PROFESSION OR TRADE OR VOCATION	Practice fully developed skills requiring intiative and judgment; keep improving at your profession or trade; becoming "tops" in your "field;" a sense of personal direction and social importance attached to your work.	Social Acceptance (status important, good citizen, full role in community life)
LIVELIHOOD	Development of philosophical approach to work, still grounded in reality but with less concern for money; having a set trade or profession assumed but used only to further learning and personal growth; discovering the "law of abundance."	Self-esteem (coming to full self-hood or individuation; sense of personal accomplishment, of being in the world)
RIGHT LIVELIHOOD	Implies a total restructuring of your role in the world; no concern for money or income; complete integration of working, playing, learning and experiencing yourself in the world.	Self-actualization, Self-fulfillment, Self-realization (full cognitive understanding of the world; peak experiences; full identification of Self and world)

RIGHT LIVELIHOOD is a conviction that what we do and how we do it should provide more than money—it should provide benefit and satisfaction to all who are touched by it.

RIGHT LIVELIHOOD is work that:

Produces something of personal and not just material benefits to others

Gives you a fair return—providing for your needs but not enough to encourage greed.

Gives you a sense of being a valued part of your community.

Develops a touchstone of deep experience by which you can measure other situations.

Gives real satisfaction.

Increases your skills and the development of all your faculties.

Gives expression to the values by which you live.

Many practitioners of Right Livelihood agree on three basic principles to use as guidelines:

1

I do it because I love doing it; I feel full when I'm doing it.

2

I find my primary reward in serving people and creating something meaningful (beautiful, useful, etc.) rather than in amassing large sums of money.

3

My relationship to my fellow beings and to my planet is based on sharing and cooperating, rather than on competing and achieving.

I suggest that when you and I take our next small step towards transforming our livelihood to Right Livelihood, then not only will our own lives be more enriched; the evolution of consciousness of our entire planet will be accelerated. We are not in this alone.

Casting Your Bread Upon the Waters: Principles of Tithing

Tithing is an ancient practice; originally, it was the tenth part of produce paid by a land tenant to the landowner. Later, as the true "owner" of all lands was recognized, the tithe was offered to God.

The percentage is not arbitrary. It is based on the numerological significance of the number ten. It is a mystical number symbolizing increase; it is the product, the completion, the seed for the next cycle. In Isaiah 6:13, it is said that if there is a tenth of normal life left in the body, it will serve as the "holy seed" of regeneration. The tenth is the concentrated part with the potential of the whole and thus symbolizes the concept that all is One. To offer a tenth is to increase one's own abundance by demonstrating faith and participation in the One, the whole system.

For purposes of making a financial commitment, the amount chosen from which to take a tenth is a matter of choice. It could be gross income, net income, or what is left after necessary expenses and obligations are met. The principle of the tenth is more important than the figure on which it is based. What should be done with the tithe? The Edgar Cayce readings suggest that we give it to causes and organizations or projects from which we ourselves have benefitted. In the broadest sense, any work which contributes to the manifestation of our highest vision of life is worthy of our support. Choose the recipients of your tithing consciously and carefully, and share your gifts with confidence and joy.

What You Can Do: Make Right Livelihood Real

The practical side of Right Livelihood is simply the move away from doing what you have to do to support yourself, and moving toward doing what you want to do and what speaks to you of its "rightness" so that supporting yourself is a natural by-product of your right effort. There are millions of people on our planet who cannot even consider this concept because of the oppression under which they struggle. You and I are not in this predicament, and it could be argued that when we relax our need to accumulate excess money and material possessions, and when we learn to labor with love, we are helping to ease the exploitive burden of the oppressed of the world.

Here are some questions which may assist you in making small, gentle moves towards a stronger position of Right Livelihood.

What am I good at?
What do I love to do?
When and where am I most creative?
When was the last occasion wherein I felt fulfilled?
What fills me with a sense of pride?
What needs doing in the world?
How may I be of real service?

Now ask yourself how much of the above is related to your present job, your work-for-pay?

Identify some ways that you can bring more fulfillment into your work, more joy, more wholesomeness.

Think about the concept of Perfection. Bring perfection to your labors. Make your labor an act of love, an act of reverence. Let your work be a vehicle for positive energy, in your own life, in those you work with, for the whole planet.

If you are not proud of what you do or what your company does, how can you change this in small and/or large ways?

To be moving toward Right Livelihood, you must come face-to-face with this reality; if your work does not allow you to express your creativity, your joy, your fullness, your own internal code of rightness, then you have three choices:

1

Use your creativity to uncover ways and means to change the system. Introduce concepts of Right Livelihood in your work place. Have patience and persistence; gently assist others to find their point of positive service.

2

Do nothing. But be honest about it. Admit to yourself that you are more invested in not rocking the boat than in changing your life, because Right Livelihood demands change for most of us, away from work-for-security, prestige and over-abundance and towards a positive, right relationship to our labor that is congruent with our roles as planetary stewards.

3

Leave this work and seek other work that has the potential of Right Livelihood. Create a situation wherein what you do, what you are good at, who you want to be, and what you feel is right are all intermingled.

It is recommended that you read *What Color is Your Parachute?* (see bibliography), not to help you find a job, but to help you identify your hidden talents and to get clear about what you love to do and how you can apply this to your present labors.

The Earthstewards Network prints a poster called Right Livelihood. Consider ordering a number of these to hang in your work environment and any environment where workers gather, to spread the awareness of this transformative concept. We live in a society, in a time and place where Right Livelihood is possible for millions of us. If we were to accept that challenge it would literally change the world.

Bibliography

Section IV (Right Livelihood)

Bolles, R. N., *What Color is Your Parachute?*
The *Foxfire* Series
Gibran, Kahlil, *The Prophet* (On Work)
Briarpatch Network, *Briarpatch Review* (see Appendix)
Briarpatch Network, *The Briarpatch Book*, New Glide Publications, 330 Ellis St., San Francisco, CA 94102
Philips, Michael, *The Seven Laws of Money*
Honest Business
Johnson, Warren, *Muddling Toward Frugality*, Shambhala, Boulder, CO
Meyer, L. E., *As You Tithe So You Prosper*, Unity School, Unity Village, MO 64065
Kamoroff, Bernard, *Small Time Operator*
Lancaster, Don, *The Incredible Secret Money Machine*
Mandino, Og, *The Greatest Salesman in the World*
Fields, Rick, *Chop Wood, Carry Water*
Hawken, Paul, *The Next Economy*
Theobald, Robert, *The Rapids of Change*
Larsen, William, *Beyond Mere Survival* (published by Lilihot Press, Box 1052, Bellingham, WA 98227)
Cameron and Elusor, *Thank God It's Monday*
Ponder, Catherine, *Open Your Mind to Prosperity*
Ross, Ruth, *Prospering Woman*
Hill, Napoleon, *Think and Grow Rich*
Gawain, Shakti, *Creative Visualization*
In Context magazine, "Living Business," Issue 11

Section V:

Sacredness of all life

Even as you do it unto the least of these you do it unto me.

—Jesus

When we respect the sacredness of all life we shall be truly free

Seeing Clearly

When we strip away our projections, our fears, and our desires and See the world clearly, what do we see? That everything is living, and that all life is sacred. Every rock, animal, human creation, natural element is alive, in that it vibrates and has its own unique Isness.

A life lived in a sacred manner is a life lived with conscious attention, respect, reverence for each human being, each inhabitant of our planet, each object. Respect means taking care with beings and things.

We don't usually think of "things" as living beings, and yet each has its own essence and its own organizing principle. At Findhorn Community in Scotland, all community-owned vehicles, pieces of equipment, etc., are given names. I remember a sign on the dashboard of a van which said something like, "I'm Doris—please shift my gears carefully and remember to check my oil." A reminder to have respect, to take care.

And how important it is to have respect for the beings for whom we have chosen to take responsibility: our children, our pets, our houseplants, the inhabitants of our gardens.

For four years, I co-existed in a small trailer with a long succession of mice, and they taught me a great deal. I learned my limits in terms of the numbers I could tolerate at one time (one!), how our territories related to each other, and, most significantly, that if I told them my requirements, they respected them. I never had to kill a mouse in that trailer.

The greatest part of really respecting each and every life is surrendering our resistance. This is true whether we are speaking of a person in our lives we don't get along with, a group of people whose ideas or behavior are repugnant to us, or ants in the kitchen. This doesn't mean letting them violate us, but rather changing on an attitudinal level from rejection to a desire to find a solution. If we need to have the other lose for us to win, then it will be difficult to learn respect and a cooperative spirit. But once we glimpse the expanded reality that can be ours if we let go of that resistance, then it's worth the hard work and soul-searching.

Some of Us Are Still With You

Frank Robson

(Reprinted from *Thinking Dolphins, Talking Whales*)

It was a beautiful day and a beautiful world.

The dolphin, when it joined us and effortlessly rode our bow wave was the final touch in a picture of peace and harmony. My daughter, Brenda, sat on the bow and dangled her feet over the edge. The dolphin came so close that she could stroke its back with her toes. Completely trusting, it swam alongside and accepted with pleasure this human contact. It was an idyllic picture with overtones of an earlier time before cruelty and greed entered into the relationship between man and animal.

Suddenly, the wickedness of what is being done overwhelmed me. These beautiful creatures taken from the sea to serve man's greed, the ocean polluted, fish and mammals diseased and dying. We no longer hunt whales with harpoons and explosives but if we continue to poison the waters in which they live they will die just as surely and far more quickly.

And the dolphins, joyous, rollicking, eager friends of humanity? Will their intelligence and sense of fun be dulled until they are good for nothing but to fall into the fisherman's net and end up in a canning factory?

I looked out towards the horizon and I thought of the great whale herds rolling past out there, busy about their breeding and feeding, harming nobody. When I spoke I was speaking to them as well as to the dolphin.

"Don't you worry, friend," I called, "some of us are still with you. They can't get the better of all of us!"

The dolphin grinned at me and swooped away sideways. I could almost have sworn it winked at me.

The Conscious Choice

John Denver

(Reprinted from the Hunger Project's *Shift in the Wind*)

Think of yourself as part of the human family, and recognize that we are all in this together. It is time for each of us to make a conscious choice. The conscious choice is to start living and reflecting that thinking: that we are part of the human family, that it is our planet, that there is enough to go around, and that we do have everything we need.

The incredible obstacles that stand in the way of the things that we've been working for so long, will disappear; because the power of the people, the power of humankind, will turn it around.

The concsious choice, in my mind, is the commitment to life, not *my* life, but Life.

Finding Our Place

Hyemeyohsts Storm

(Reprinted from *Seven Arrows*)

Our Teachers tell us that all things within this Universe Wheel know of their Harmony with every other thing, and know how to *Give-Away* one to the other, except man. Of all the Universe's creatures, it is we alone who do not begin our lives with knowledge of this great Harmony.

All the things of the Universe Wheel have spirit and life, including the rivers, rocks, earth, sky, plants and animals. But it is only man, of all the Beings on the Wheel, who is a determiner. Our determining spirit can be made whole only through the learning of our harmony with all our brothers and sisters, and with all the other spirits of the Universe. To do this we must learn to seek and to perceive. We must do this to find our place within the Medicine Wheel. To determine this place we must learn to *Give-Away*.

Learning

T. H. White (Reprinted from *The Book of Merlyn*)

"The sentries," he enquired. "Are we at war?"

She did not understand.

"War?"

"Are we fighting against people?"

"Fighting?" she asked doubtfully. "The men fight sometimes, about their wives and that. Of course there is no bloodshed, only scuffling to find the better man. Is that what you mean?"

"No. I meant fighting against armies: against other geese, for instance."

She was amused at this.

"How ridiculous! You mean a lot of geese all scuffling at the same time. It would be amusing to watch."

Her tone surprised him.

"Amusing to watch them kill each other!"

"To kill each other? An army of geese to kill each other?"

She began to understand the idea very slowly and doubtfully, an expression of grief and distaste coming over her face. When it had sunk in, she left him. She went away to another part of the field in silence. He followed her, but she turned her back. Moving round to get a glimpse of her eyes, he was startled by their abhorrence: a look as if he had made an obscene suggestion.

He said lamely: "I am sorry. You do not understand."

"Leave talking about it."

"I am sorry."

Later he added: "A person can ask, I suppose. It seems a natural question, with the sentries."

But she was thoroughly angry, almost tearful.

"Will you stop about it at once! What a horrible mind you must have! You have no right to say such things. And of course there are sentries. There are the jerfalcons and the peregrines, are there not: the foxes and the ermines and the humans with their nets? These are natural enemies. But what creature could be so low and treacherous as to murder the people of its blood?" . . .

Out loud, he said: "I was trying to learn."

What You Can Do: Increasing Awareness of the Sacredness of All Life

Here is a collection of suggestions as to how you might make the concept of sacredness of all life more real to yourself and others. Thanks to Nancy Binzen, Corinne McLaughlin, Irv Thomas, Margaret Boarman, and Ed Setchko for valuable contributions to this section.

Celebrate the important milestones of life, such as birth, death, puberty, adulthood, marriage, divorce (and other important beginnings and endings).

Make friends with a spider.

Before each meal, take time to focus on and appreciate the life that has given itself for your sustenance. Also be aware of life given for your warmth, comfort, adornment, etc.

Understand the obtuse behavior of a Russian, a military man, a rapist, a . . . (name your own villain).

Spend time with nature regularly; just sit quietly and watch.

Recognize your own value, your sacredness; then you will see the sacredness of all.

Explore ways of dealing with household "pests," other than killing them.

Explore ways of dealing with societal "pests," other than killing them.

Support efforts to limit animal experimentation.

Support Greenpeace, Save the Whales, dolphin research, other groups who care about other life forms.

Join the Hunger Project.

Notice when you feel hostile toward an individual, a group, a country; are the traits you reject in them hiding in the shadows of your own personality?

Plant a tree.

Volunteer your time in a spay clinic.

Identify that which needs doing in a convalescent hospital in your area.

Adopt a refugee child or family.

Work for responsible cutting of timber and re-foresting; find out about the Earthstewards Network's Peace Trees projects.

Learn about eco-systems and how each tiny part plays its unique and important role.

Notice how you and others speak to and act toward children, pets, plants.

Teach your children to respect plants, animals, people. (Remember that children learn by example.)

Find out about hospice programs in your area.

Join a citizen diplomacy group to the Middle East.

Learn the culture, history and language of those we fear as different (e.g., communist block nations).

Attend a Warriors of the Heart conflict-resolution training sponsored by the Earthstewards Network.

Bibliography

Section V (Sacredness of All Life)

Watson, Lyall, *Supernature*
Gifts of Unknown Things
Lifetide
Hunger Project, *A Shift in the Wind* (news report) see Appendix
The Brandt Commission Report, MIT Press, Cambridge, MA (the report from the Commission headed by Willy Brandt of Germany, detailing the reasons for hunger and poverty and inequitable distribution of essentials; an important document)
Lilly, John, *Man and the Dolphins*
Snyder, Gary, *Turtle Island*, other works
Bergon, F., ed., *The Wilderness Reader*
Dillard, Annie, *Pilgrim at Tinker Creek*
Castaneda, Carlos, all his works
Storm, Hyemeyohsts, *Seven Arrows*
Publications from Greenpeace and Friends of the Earth
Robson, Frank, *Thinking Dolphins, Talking Whales*
White, T. H., *The Book of Merlyn*
Lappé, Frances Moore and Collins, Joseph, *Food First: Beyond the Myth of Scarcity*
Harding, Esther, *Woman's Mysteries*
Kubler-Ross, Elizabeth, *On Death and Dying*
Grollman, E. A., *Talking About Death*
Trumbo, Dalton, *Johnny Got His Gun*
Institute for Soviet-American Relations, *Surviving Together*
Holyearth Fdtn., *Just People, a Handbook for and about Citizen Diplomats*
Holyearth Fdtn., *Russian Words, Phrases and Travel Tips* (60 min. tape)
Macy, Joanna, *Despair and Personal Power in the Nuclear Age*
Hawkes, Glenn, *What About the Children?*
Carlson and Comstock, *Securing Our Planet, Citizen Summitry*
Werner and Schuman, *Citizen Diplomats*
Wright, M.S., *Behaving As if the God in All Life Mattered*
Fuller, Buckminster, *Utopia or Oblivion*
Fromm, Erich, *Art of Loving*

Section VI:

EGO TRANSCENDENCE

. . .and having found one pearl of great price, he went and sold all that he had, and bought it.

—Jesus

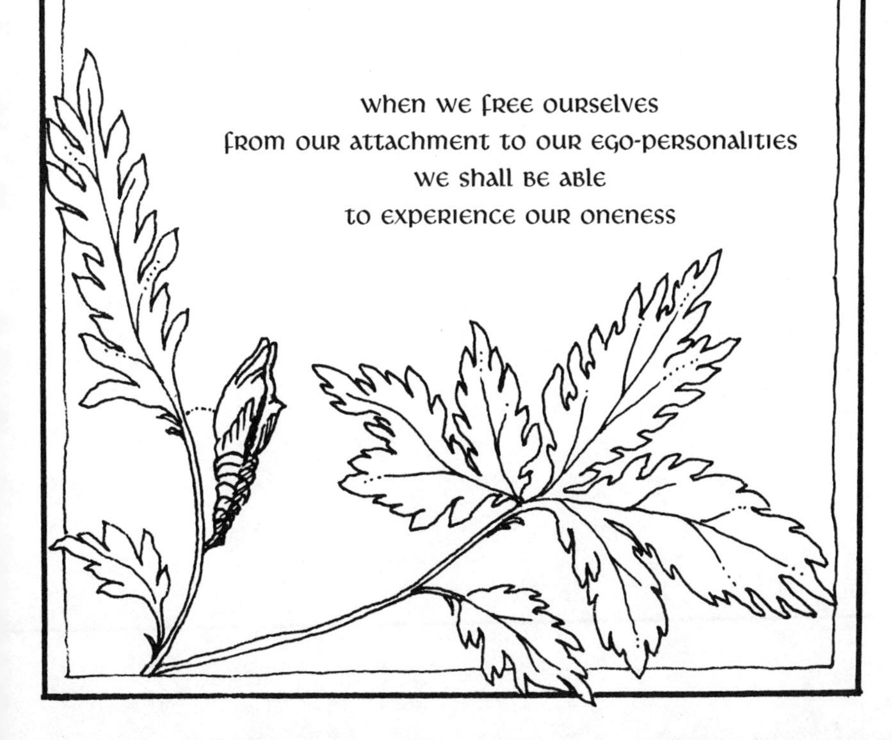

when we free ourselves
from our attachment to our ego-personalities
we shall be able
to experience our oneness

Beyond the Drama

Do not kill your personality; do not annihilate your personality in the sense of wiping it out. You have brought it into being yourself; it is part of you, the emotional and psychical part of you . . . the evolutionary work of aeons upon aeons in the past. Raise the personality. Cleanse it, train it, make it shapely and symmetrical to your will and to your thought, discipline it, make it the temple of a living god so that it shall become a fit vehicle, a clean and pure channel for passing into the human consciousness the rays of glory streaming from the god within . . . from the spiritual or divine consciousness.

G. dePurucker
(Reprinted from *Golden Precepts of Esotericism*)

The path to freedom is the one that leads beyond attachment. The desert fathers spoke of it as "to be in this world yet not of this world." Most of us spend most of our lives totally caught up in the drama of everyday reality, with its ups and downs, its depressions and its ecstasies, its losses and its gains. We become addicted to pleasure, which makes us targets for pain; we become attached to success, which opens us to failure. Our mood swings toss us erratically from the mountain to the valley and back again.
We are prisoners of our personalities.

Happiness has never been reached by seeking it; happiness is a by-product of living one's life in harmony with one's purpose for being. Your personality is a set of garments that you wear, to clothe

the true you appropriately in a particular set of situations. When you forget that you *have* personality and think that you *are* your personality, then your ego, which is the controlling function of your personality, takes over the control for your whole being. This creates an imbalance, and you become susceptible to image, to approval, to manipulation, as you try (unsuccessfully) to "be who they want me to be." As long as you think that you are your personality, then the locus of control of your life will be external to you, not internal as it was meant to be. The ego is a superb servant and a terrible master.

How to reclaim control of your being in the world? How to return the ego to its proper function in your life? The first step is to understand truly that you *have* an ego and a personality, and that you *are* he or she who manifests your part in the plan through these exquisite vehicles. An effective next step is that of non-attachment. To quote Ram Dass' oft-repeated aphorism:

Totally involved
Totally unattached

The ego gains control through attachment, clinging, hanging on—through addictions. When you are no longer addicted to anything or anyone, then the ego has nothing to work with, and it surrenders. But it does not surrender gracefully.

Being totally involved, fully present to each moment and yet not "hooked" to any of it may seem impossible, but not if taken in small steps. A good model for understanding this is presented in

Ken Keyes' *Handbook to Higher Consciousness.* It uses the concepts of addictions, expectations, and preferences.

The world of addictions: this is the polar opposite of non-attachment. You are holding on to something or someone addictively when you feel that it is beyond your control to decide whether or not to continue the connection.

The world of expectations: a less-trapped state wherein you have very strong biases and demands about how he/she/the universe should behave. You have an expectation when you experience emotional discomfort over things not going your way.

The world of preferences: a more centered state wherein you have opinions as to how you wish the universe to behave, but you can allow for outcomes other than those that meet your view of how it is.

Now the path away from a self-created emotional and psychic prison is obviously away from addictions, a process of upleveling your addictions to expectations and your expectations to preferences. This returns the locus of control from outside yourself to inside yourself.

Consider spending some time with your journal, listing all of your addictions, expectations, and preferences. Everything from political opinions to coffee to sex to surrender itself. (Are you addicted to surrender?)

So now what? Uplevel! Change your addictions to expectations and your expectations to preferences. Look at your life. What are you holding onto? What are you so attached to that just thinking about letting it/her/him go gives you hives? There's something to work with.

Begin now to turn that addiction into an expectation and then into a preference. This is the path toward freedom, for yourself and for everyone to whom you relate.

Some Potent Words

If the desire to be honest is greater than the desire to be "good" or "bad," then the terrific power of one's vices will become clear. And behind the vice the old forgotten fear will come up (the fear of being excluded from life) and behind the fear the pain (the pain of not being loved) and behind this pain of loneliness the deepest and most profound and most hidden of all human desires: the desire to love and to give oneself in love and to be part of the living stream we call brotherhood. And the moment love is discovered behind hatred all hatred disappears.

Fritz Kunkel, M.D., 1899–1956. American psychiatrist.
(Reprinted from *In Search of Maturity*)

Heaven is lasting and earth enduring.
The reason why they are lasting and enduring is that they do not live for themselves;
Therefore they live long.
In the same way the Sage keeps himself behind and he is in the front;
He forgets himself and he is preserved.
Is it not because he is not self-interested
That his self-interest is established?

Tao Te Ching

I had gone a-begging from door to door in the village path, when thy golden chariot appeared in the distance like a gorgeous dream and I wondered who was this King of all kings!

My hopes rose high and methought my evil days were at an end, and I stood waiting for alms to be given unasked and for wealth scattered on all sides in the dust.

The chariot stopped where I stood. Thy glance fell on me and thou camest down with a smile. I felt that the luck of my life had come at last. Then of a sudden thou didst hold out thy right hand and say "What hast thou to give to me?"

Ah, what a kingly jest was it to open thy palm to a beggar to beg! I was confused and stood undecided, and then from my wallet I slowly took out the least little grain of corn and gave it to thee.

But how great my surprise when at the day's end I emptied my bag on the floor to find a least little grain of gold among the poor heap. I bitterly wept and wished that I had had the heart to give thee my all.

Rabindranath Tagore
(Reprinted from *Gitanjali*)

What You Can Do: A Worksheet for Ego Transcendence

The path that takes us beyond the dance of the ego goes right through the dance. We must experience this ego to move beyond it. There is no way of making an end-run around it. Following are a few reflections that may assist us in getting "unstuck" as we move along the path. This is but a sample of the multitude of vehicles we can use to open up our awareness of how we limit our limitlessness. Make use of any vehicle that works for you.

I.

A. Make a list of those people that you dislike, hate, mistrust, fear.

B. Identify the characteristics, qualities, behaviors of these people that generate your negative feelings.

C. You have now identified many of the attributes within yourself that you have buried in your Shadow. You must learn to accept these attributes in your own being before you can move beyond them. The big first step toward accepting yourself fully is this initial awareness of all of the parts of you.

II. Fill in the blanks.

A. I usually act ______________________________

because ______________________________.

B. I usually feel ______________________________

because ______________________________.

C. I usually think ______________________________

because ______________________________.

D. I usually get attention from my friends when I __________

because I ______________________________

and they ______________________________.

E. When I don't get my way I usually ______________________

because ______________________________

and I feel ______________________________.

F. When people are angry with me, I ______________________

because ______________________________

and then I ______________________________.

G. Whenever I'm upset with someone I usually ____________

because ______________________________

and then I feel ______________________________.

H. My mother always used to say that I ________________

and now I ______________________________.

I. What I like best about myself is ______________________

because ____.

J. What I like least about myself is ______________________

because ____.

K. My epitaph will read __________________________________

because ____.

Read over your answers, and perhaps write about your feelings in your journal. You are involved in a process of knowing yourself more fully. Don't get too heavy, too attached. Just let it flow out of you and then move on.

III. Where are my stuck places? . . . with regard to:

A. Security ____

B. Possessiveness ______________________________________

C. Sexuality ____

D. Emotions ____

E. Power and Control ___________________________________

F. Concern for others ___________________________________

G. Opening my heart ____________________________________

H. Communicating my feelings ______________________

I. Acknowledging and using my talents ______________

J. Spiritual awakening ____________________________

K. Oneness/my connection to God __________________

What will be my next step toward getting unstuck in each of the above areas?

IV.

Think of your journey away from mediocrity and toward a conscious, full, actualized life. Identify three areas of your personal growth, as follows:

A. An issue or area that is in process, that you are handling quite well. It's working out OK.

B. A growth issue or area that is HOT! This is your #1 priority now; front burner.

C. An issue or area that you know is there, but you're not quite ready to tackle just yet; back burner. You'll deal with this one some day.

Simply identify and describe these three growth areas in your journal. Nothing else to do. Remember, the name of the game is awareness, not solutions. Once we allow ourselves to be fully aware of our process, we activate our "automatic pilot," which faithfully guides us on our correct path.

Bibliography
Section VI (Ego Transcendence)

Kunkel, Fritz, *In Search of Maturity*
Ram Dass, *Be Here Now*
Grist for the Mill
The Only Dance There Is
A Course in Miracles (Foundation for Inner Peace)
All the Edgar Cayce material (send for book and tape catalogue to A.R.E.; see Appendix)
Jung, Carl, *Memories, Dreams, Reflections*
Man and His Symbols
Yogananda, Paramahansa, *Autobiography of a Yogi*
Prather, Hugh, *Notes to Myself* and *How To Live in This World and Be Happy*
Philips, Howes, & Nixon, eds., *The Choice Is Always Ours*
Golas, Thaddeus, *The Lazy Man's Guide to Enlightenment*
Rajneesh, Bhagwan Shree, *The Mustard Seed* (an inspiring and challenging commentary on the Gospel of Thomas by a non-Christian wise man)
Ferguson, Marilyn, *Aquarian Conspiracy* (a must for all who feel themselves part of the transformation of our planet!)
Smith, Adam, *Powers of Mind*
Stevens, Barry, *Don't Push the River*
Rogers, Carl, *On Becoming a Person*, many other works
Sheehy, Gail, *Passages*
Huxley, Laura, *You Are Not the Target*
Shah, Idries, *Tales of the Dervishes*
Nasrudin
Keyes, Ken, *Handbook to Higher Consciousness*
Prescriptions for Happiness
Kazantzakis, Nikos, *The Last Temptation of Christ*
The Tao Te Ching (any version)
Tagore, Rabindranath, *Gitanjali*
dePurucker, *Golden Precepts of Esotericism*
Levine, Stephen, *Who Dies?* and *Gradual Awakening*
Gawain, Shakti, *Living in the Light*
Ury, Bill and Roger Fisher, *Getting to Yes*

Section VII:

Oneness

Before the beginning of time
Withstanding the solitude of our unity
We dared to part
To gamble
To lose
What could not be lost.

(Reprinted from *Seed*)

When we experience our Oneness—
our total connectedness
with all beings
we shall be at peace
within our own hearts

There is But One Human Heart

Can you feel it?
The rich red blood
your heart sends coursing
through your human body?

Can you feel it?
This same rich blood of life
being pumped by every human heart
in every mountain, valley, plain,
on our common home—the earth?

From the sea, from the caves,
from the death-point in the battle sphere,
we have struggled and survived.
We have borne one another and killed one another
and we have, in some incredible way,
passed onward the tenuous thread
of our common human life.

We have a history, you and I.
We are the human ones.
We have come to learn from and move beyond
our arrogance and our separateness.
We have come here to go home.
Together

The political boundaries, the walls we have made,
they will crumble.
Our common blood will remain.
We will come together as a One World Family,
even as you and I now come together
to challenge the walls we have built between us.
We will share our love for all our family
just as you and I now reach out
to touch the place of oneness within our common heart.

Danaan Parry

Oneness

At the very heart of the Sevenfold Path of Peace is the awareness of our Oneness. Our Oneness, our total connectedness with all beings, is one-half of the paradox that wise ones have wrestled with for ages. The other half of the paradox is our Separateness. Each of us is, at the same moment, totally one with all and also a completely separate, unique being. Because we have trained our intellects to carry almost all of the burden of knowing, it is easy to comprehend the separateness pole of the paradox. Our five senses report to us, "I am me and you are you and that tree is a tree and Richard Nixon is—well, he's definitely not me!" To comprehend the other side of the coin, which is equally true at each instant, namely that you and I and the tree and—yes—him, too, are one consciousness, we must move beyond the limitations of our five physical senses and allow different tools to work for us. In addition to our intellects, which are exquisite tools within the arena of rational, linear, causal processes, we also have a full set of intuitive tools. These have lain dormant for most of us, but they can be rediscovered and used to assist us in our exploration of the other half of our existence: the non-linear (mosaic), acausal world which humankind now stands on the brink of discovering (again). Quantum physics is beginning to reaffirm, in its wonderful western methodical manner, the ancient Taoist, Buddhist, and Vedic wisdom that has told us of our Oneness for thousands of years. The Rig Vedas say that if you and I and everyone knew who we really were, then there would be only one of us looking at ourself. Jesus said, "I and the Father are one," and Zen Buddhism tells us that "separateness is an illusion." And now from quantum physics, Bell's Theorem, which helps us to understand how our universe works, says that nothing can occur in the universe that does not significantly affect every other thing in the universe. If a star explodes in a galaxy a million light years away, if a dog dies in China, you are changed, not only psychically, but also within every molecule of your being. The Upanishads say, "If anyone in the world is hungry, you are hungry. If anyone is suffering, you are suffering." This is literally true; we are One.

It is not necessary that we "understand" our Oneness, for understanding is the way of the intellect. Rather, we can *feel* this intimate, total connection, experience it with our hearts, allow our inner voices to whisper to us of this oneness of all life. This experience of Oneness is reflected in the eastern greeting, "Namasté." Namasté means, "The place within me where God dwells honors the place within you where God dwells," or, "That which is God in me sees that which is God in you." Just imagine what would happen in our lives if we were able to see that which is God within each person we meet, beyond the images, beyond the rough, sometimes threatening exterior.

There is a model of Oneness from Edgar Cayce that helps me to see my path toward the experience of this truth. The journey is in moving beyond the simple awareness of my five senses, my conscious mind and rational thought, and to a familiarity with my unconscious mind, which is the area of my dreams, my cellular wisdom and inner guidance, and finally to an awareness of my superconscious mind, which is the place of total knowing. At that place I become aware of what my own personal superconscious mind really is: it is the total connection between all beings. It is my divine nature. My Superconsciousness is your Superconsciousness is everybody's Superconsciousness.

This is our point of Oneness, in which I can become aware of who I really am and why I am really here.

Namasté.

Danaan Parry

Many Voices, One Wisdom

The Great Heresy and the only real heresy is the idea that anything is separate, distinct, and different essentially, from other things. That is a wandering from natural fact and law, for Nature is nothing if not co-ordination, co-operation, mutual helpfulness; and the rule of fundamental unity is perfectly universal: everything in the Universe lives for everything else.

G. dePurucker
(Reprinted from *Golden Precepts of Esotericism*)

Thou art One, the first of every number, and the foundation of every structure.
Thou art One, and at the mystery of Thy Oneness the wise of heart are struck dumb,
For they know not what it is.
Thou art One, and Thy Oneness can neither be increased nor lessened;
It lacketh naught, nor doth aught remain over.
Thou art One, but not like a unit to be grasped or counted,
For number and change cannot reach Thee.
Thou art not to be envisaged, nor to be figured thus and thus. . . .

Unknown

If everyone in this room knew who they really were, there would be only One of us in this room.

Ram Dass

I and the Father are One.

Jesus

What You Can Do: Moving Toward (W)holism

Oneness is not an intellectual concept that can be debated, proven or disproven. It simply is. It is our state of being in its most basic form. Here are some ideas that may help you to invite the experience of Oneness into your awareness.Try a good number of these while holding the vision of our Oneness in your mind's eye.

Om for world peace. (By yourself or in a small group, join a worldwide network of people who chant Om every Sunday night, focusing on world peace and the healing of our planet.)

Get involved with the Earthstewards Network's Middle East citizen diplomacy program.

Pray consciously for the realization of Oneness.

Ride a city bus with full awareness of the concept of the divine in all beings.

Participate in an evening of singing with Heavensong, Box 450-C, Kula, Maui, Hawaii 96790

Do the Dances of Universal Peace (Sufi dancing).

Touch people when you talk to them.

Give a massage to a very old person or a very young one.

An exercise for two people: sit facing each other and look silently into each other's eyes without wavering your gaze. Don't stare; just look. (It helps to look at the bridge of the nose between the eyes.) Do this for five minutes and observe the results.

Take a self- or guided hypnosis journey to your point of Oneness with all.

Learn the ways that you stereotype people; take a group in understanding your subtle prejudices.

Support the United Nations.

Visit all types of churches.

Smile every day at someone who is "different."

Join Planetary Citizens (see Appendix).

Work for integration (on many levels!)

Ferret out the good points in those you mistrust.

Make a new friend weekly.

Chant in a circle of friends for a long time and experience your Oneness.

Bibliography

Section VII (Oneness)

Spangler, David, *Toward a Planetary Vision*
de Chardin, Teilhard, *The Phenomenon of Man*
Capra, Fritjof, *The Tao of Physics*
Leonard, George, *The Silent Pulse*
Bentov, I., *Stalking the Wild Pendulum*
Boulden, Jim, *We Are One*
Ferguson, Marilyn, *The Aquarian Conspiracy*
Ram Dass, *The Only Dance There Is*
The Bible
The Edgar Cayce material (see A.R.E. in Appendix)
Sun Bear & Wabum, *The Medicine Wheel*
Kubler-Ross, Elizabeth, *On Death and Dying*
Metzner, *Maps of Consciousness*
Al Huang, *Living Tao*
Purce, Jill, *The Mystic Spiral*
Buber, Martin, *I and Thou*
Heinlein, *Stranger in a Strange Land*
Campbell, Joseph, *Masks of God*
Watts, Alan, *Man, Woman, and Nature*
dePurucker, *Golden Precepts of Esotericism*
Seed, Harmony Press
Zukav, Gary, *The Dancing Wu Li Masters*
Spiritual Community Guide
Taylor, Kathryn, *Miscellany* (Seed Thoughts); 10 Pleasant Ln., San Rafael, CA 94901
Whole Life Times (bi-monthly) Suite 1, 132 Adams St., Newton, MA 02146
Common Ground (listing of classes, events, etc.) 1300 Sanchez St., San Francisco, CA 94131
Neumann, Erich, *The Great Mother*, Princeton University Press
Swimm, Bryan, *The Universe is a Green Dragon*

Bibliography

Appendix
Activities and Organizations

The groups listed here offer a wide variety of services to individuals and to our planet. Needless to say, there are hundreds more available, and we invite you to share them with us by sending a description and literature on them to the Earthstewards Network. We hope you will make some satisfying new connections in browsing through the list and we look forward to hearing from you.

Abundant Life Seed Foundation
PO Box 772, Port Townsend, WA 98368
(206) 385-5660

A hardworking group dedicated to preserving and offering vanishing varieties of untreated vegetable, culinary and medicinal herbs, flowers, and tree seeds. They also sponsor the World Seed Fund, a hand-to-hand project giving free seeds to people that need them, here and throughout the world.

American Rivers
801 Pennsylvania Ave. SE, Suite 400, Washington, DC 20003
(202) 547-6900

Works to protect a priority list of wild, natural, and free-flowing rivers and their landscapes and acts as an information resource for river activists. Works on legislation and assists local and state organizations in conservation projects. Newsletter updates members on river-related legislative action, provides contact names, and lists recreational opportunities.

Americans for the Environment
1400 16th St. NW, Washington, DC 20036
(202) 797-6665

Teaches people how to use political campaigns and elections to promote environmental protection. Trains groups and activists across the nation in campaign-related skills, such as media relations, volunteer recruitment, and fundraising.

Amnesty International USA
322 Eighth Avenue, New York, NY 10001
(212) 807-8400

An independent worldwide movement working impartially for the release of all prisoners of conscience, fair and prompt trials for political prisoners and an end to torture and executions.

Angela Center
535 Angela Drive, Santa Rosa, CA 95403
(707) 528-8578

An adult education center and retreat operated by the Ursuline nuns. Seeks to integrate spirituality, psychology, social responsibility and the arts, believing that religious experience can best be understood through its expression in our daily lives.

Association for Humanistic Psychology
1772 Vallejo Street, San Francisco, CA 94103
(415) 346-7929

World-wide network for the development of the human sciences recognizing our distinctly human qualities, working to fulfill our innate capabilities as individuals and as members of society.

Association for Research and Enlightenment, Inc.
PO Box 595, 67th Street and Atlantic Ave., Virginia Beach, VA 23451
(804) 428-3588

A living network of people who are finding a deeper meaning in life though the psychic work of Edgar Cayce. On the leading edge in such areas as holistic health care, meditation instruction, reincarnation studies and spiritual healing. Publications, field seminars, Study Groups and monthly membership mailings.

Beyond War
222 High Street, Palo Alto, CA 94301
(415) 328-7756

A nonprofit, non-partisan foundation whose goal is to educate through public presentations, writings, group discussion, and audio-visual techniques, that in the nuclear age, war is obsolete as a means of resolving conflict; and, to research and develop educational processes through which an individual can discover and become personally involved, working in cooperation with people of all nations, races, and religions, to build a world beyond war.

Brain/Mind Bulletin
PO Box 42211, Los Angeles, CA 90042-4406
(213) 223-2500

A monthly report edited and published by Marilyn Ferguson, best-selling author of the *Aquarian Conspiracy*. Discusses newly discovered possibilities in brain science, creativity, psychology, education, learning research and studies of states of consciousness.

Breitenbush Community
PO Box 578, Detroit, OR 97342
(503) 854-3314

A restored hot springs resort and community that offers a place for personal retreats and a wide variety of conferences and workshops, including massage and meditation.

Center for U.S./U.S.S.R. Initiatives
3268 Sacramento Street, San Francisco, CA 94115
(415) 346-1875

A non-profit, citizen-initiated organization dedicated to improving the quality of communication between the United States and the Soviet Union, to reducing US-USSR hostility by bringing Soviet and American peoples together in non-political, trust-building encounters and projects, and to providing updated, balanced public education on US-USSR issues to a broad spectrum of American citizens.

Children Around the World Resource Center
PO Box 40657, Bellevue WA 98007
(206) 643-0172

Global education, resources and pen-pals for children in grades K thru 9. children's global art collection.

Chinook Learning Center
Box 57, Clinton, WA 98236
(206) 321-1884

A non-profit educational center and community dedicated to exploring a comprehensive vision of the future through programs focusing on personal, regional and global issues. Their perspective is spiritually-based, especially inspired by humanity's new relationship with the earth. Offers a variety of workshops, conferences and long-term programs to help people develop the understanding and the skills to effect positive change in themselves and in the world.

Citizen's Clearinghouse for Hazardous Wastes
PO Box 6806, Fall's Church, VA 22040
(703) 276-7070

Teaches individuals and grassroots organizations how to mobilize against local environmental hazards. Science department helps people interpret scientific information. Provides strategic and organizational advice as well as training in such direct-action techniques as protesting, blockades, and demonstrations.

Clamshell Alliance
PO Box 734, Concord, NH 03301
(603) 224-4163

Works to stop construction of nuclear energy plants in New England, particularly Seabrook in New Hampshire. Meetings, educational forums, lending library with videos and books, peaceful protests, and canvassing with educational materials are part of its program.

Clean Water Action Project
1320 18th St., NW, 3rd Floor, Washington, DC 20036
(202) 457-1286

Works for clean and safe water at an affordable cost, control of toxic chemicals, and the protection of natural resources. Techniques include door-to-door canvassing, petition drives, and meetings.

Concerned Educators Allied for a Safe Environment
17 Gerry St., Cambridge, MA 02138
(617) 864-0999

Addresses issues of pollution, nuclear power, nuclear war, and the draining of social welfare resources by the military budget. Membership comprises early childhood teachers, parents, and others concerned about preserving a healthy environment for children to grow up in.

Connect/US-USSR
2525 E. Franklin Ave., Minneapolis, MN 55406
(612) 333-1962

A non-profit organization working to foster educational and cultural exchanges between the United States and the Soviet Union. Includes professional exchanges, the hosting of Soviet visitors, and the facilitation of educational and cultural programs. The purpose of these connections is to educate Americans and Soviets about one another and to build mutually beneficial relationships between the people of the two countries.

Defenders of Wildlife
1244 19th St. NW, Washington, DC 20036
(202) 659-9510

Uses lobbying, public education, litigation, and a network of citizen activists to help protect wildlife and their habitats, strengthen the Endangered Species Act, and preserve natural sanctuaries. Publishes the bimonthly magazine *Defenders.*

Earth First!
PO Box 5176, Missoula, MT 59806-5716
(406) 728-8114

Works to preserve wilderness and natural diversity and to advance a biocentric philosophy. Those involved in the Earth First! movement engage in civil disobedience, education, wilderness proposals, and letter writing. A subscription to *Earth First!* costs $20.

Earthstewards Network
PO Box 10697, Bainbridge Island WA 98110

Personal and global expansion of consciousness, spiritual awareness for creation of postitive alternatives; workshops, publications, tapes, trips, posters, newsletter and journal.

Environmental Action
1525 New Hampshire Ave., NW, Washington, DC 20036
(202) 745-4870

Works to stop the contamination of the environment by toxic wastes, find safe and affordable energy alternatives, and promote recycling and reduction of solid waste through research, legislation, political action, public education, and three bimonthly magazines, *Environmental Action, Powerline,* and *Wasteline.*

Environmental Defense Fund
257 Park Ave. S, New York, NY 10010
(212) 505-2100

Dedicated to generating innovative solutions to environmental problems such as acid rain, toxic wastes, and nuclear energy. Members' financial contributions help fund the development of legislation, the creation of regulatory programs, and the strengthening of local environmental groups.

Environment Policy Institute/Friends of the Earth
218 D St. SE, Washington, DC 20003
(202) 544-2600

Engaged in research, public education, legislation, and lobbying to promote energy and water conservation; clean air, water, and groundwater; protection of farmland, food resources, and public health; reduction of US dependence on nuclear power and foreign oil; the reclamation of strip-mined lands; and international water projects to meet development goals without damaging the environment.

Esalen Institute
Big Sur, CA 93920
(408) 667-3000

A center to explore those trends in education, religion, philosophy, and the physical and behavioral sciences which emphasize the potentialities and values of human existence. Its activities consist of seminars and workshops, residential programs, consulting and research.

The Fellowship of Reconciliation
Box 271, Nyack, NY 10960
(914) 358-4601

A group of women and men who recognize the essential unity of all humanity and have joined together to explore the power of love and truth for resolving human conflict. Committed to the achievement of a just and peaceful world community, with full dignity and freedom for every human being. Seeks the company of people of faith who will respond to conflict nonviolently, seeking reconciliation through compassionate action.

Findhorn Community
Forres IV 36 OTZ, Scotland

Global consciousness, spiritual evolution, community; guest programs, books, tapes, *One Earth* journal.

Food First/Institute for Food & Development Policy
145 Ninth Street, San Francisco, CA 94103
(415) 864-8555

Research and education on world hunger and nutrition (founded by Frances Moore Lappe); education projects, pamphlets, studies, newsletter.

Friends of the Earth
377 City Road, London ECIV-INA, United Kingdom.

Ecology action, research, education; publications.

Gesundheit Institute
2630 Robert Walker PL., Arlington, VA 22207
(703) 522-0970

Trying to build a free interdisciplinary health care community for all people. Based on a 12-year experiment of giving free health care (not accepting insurance - not carrying malpractice insurance) and living with patients as friends.

Global Action Plan for the Earth
449 A Route 28A, West Hurley, NY12491
(914) 331-1312

A new organization founded by David Gershon to help individuals work on environmental goals in their communities through the establishment of a Household EcoTeam Program.

Global Tomorrow Coalition
1325 G. St. NW, Suite 915, Washington, DC 20005-3104
(202) 628-4016

Organizes task forces on the community and national levels to discuss environmental issues and to work toward passing legislation affecting the future of the world's natural resources. Addresses global trends in population growth, wasteful resource consumption, environment degradation, and sustainable development.

Greenpeace
1436 U St., NW, Washington, DC 20009
(202) 462-1177

With 25 million supporters and offices in 18 countries, Greenpeace works for no nukes, no toxics, ocean ecology, wildlife preservation—all with a view toward generating media attention. Greeenpeace was the first high-profile environmental group to move beyond the safe confines of lobbying and electoral work to engage in direct action. The most effective users of bumper sitckers as a political tool in history.

Hunger Project
1388 Sutter St., San Francisco, CA 94109
(415) 928-8700

Raising planetary consciousness to the possibility of ending hunger; fund-raising, education.

In Context Institute
Box 11470, Bainbridge Island, WA 98110-5470
(206) 842-0216

Devoted to assisting in the development of a humane and sustainable world through citizen diplomacy and environmental programs. Quarterly magazine is $18/year.

Institute of Cultural Affairs
4750 North Sheridan Road, Chicago, IL 60640
(312) 769-6363

A private, not-for-profit, research training and demonstration group concerned with the human factor in world development. Committed to serve in the process of improving the quality of human life through participatory problem-solving techniques, curriculum designs, educational methods and conferences. Focus is on helping people help themselves.

Institute for Soviet-American Relations
1601 Connecticut Avenue, Suite 301, NW, Washington, DC 20009
(202) 387-3034

Makes connections between people, ideas and systems to help weave the fabric of Soviet-American relations. Publishes *Surviving Together: A Journal on Soviet-American Relations*, reflecting the good news about Soviet-American relations—exchanges, public education programs, current affairs in the USSR, official contacts, etc. Also publishes a *Handbook of Organizations Involved in Soviet-American Relations.*

Kripalu Center
PO Box 793, Lenox, MA 01240
(413) 637-3280

Located in an exquisite Berkshire mountain setting, this is a residential community of 250 offering year-round programs in yoga, personal growth, spiritual attunement, holistic health, fitness and yoga teacher training. Two or four month spiritual lifestyle training available where you work with the residents. Books and audio and video tapes available through catalog. Free program guide.

National Audubon Society
950 Third Ave., New York, NY 10022
(212) 832-3200

Founded in 1905, this organization is regarded as one of the more conservative, mainline conservationist groups. With a membership exceeding a half million, the National Audubon Society not only maintains a quarter million acres of wildlife sanctuary, but it also is responsible for publishing glossy magazines like *Audubon* and *American Birds.*

Nature Conservancy
1815 N. Lynn St., Arlington, VA 22209
(703) 841-5300

Purchases ecologically significant areas to ensure their preservation. Identifies and locates rare and endangered species, protects habitats through land acquisition by gift or purchase, and manages more than one thousand preserves. Encourages use of sanctuaries by researchers, students and the public.

New Age Society
342 Western Avenue, Brighton, MA 02135
(617) 787-2005

An information resource for people interested in pursuing New Age thinking, values and lifestyle. Membership includes a subscription to *New Age* magazine, an annual resource directory, book discounts and other New Age services.

New Alchemy Institute
237 Hatchville Rd., E. Falmouth, MA 02356
(508) 564-6301

Research and education in appropriate technology, agriculture, aquaculture, tours of projects, workshops, consultation, yearly journal.

New Options
Box 19324, Washington, DC 20036
(202) 745-7460

A magazine that sensitively asks the embarrassing questions that must be asked for our new movements to become responsible and effective. Covers national and international news.

Oxfam America
115 Broadway, Boston, MA 02116
(617) 482-1211

A non-profit, international agency that funds self-help development projects and disaster relief in poor countries in Africa, Asia, and Latin America, and also prepares and distributes educational materials for Americans on issues of development and hunger.

Peace Child Foundation
9502 Lee Highway, Fairfax, VA 22031
(703) 385-4494

Producers of a peace play enacted by children of the world. The lively story deals with the nations in conflict and how children lead the world to peace.

Peace Links
747 8th Street, SE, Washington, DC 20003
(202) 544-0805

A nationwide network of women who are reaching and activating an entirely new constituency of people who have made a commitment to preventing nuclear war. Includes personal and political action, educational materials and events.

Peace People
Fredheim, 224 Lisburn Road, Belfast BT96GE, Northern Ireland

A newsletter about the peace efforts in Northern Ireland. A good way to get a perspective on this troubled land.

Peace Table
9712 Edmonds Way, #162, Edmonds, WA 98020

Peacemaking through culinary diplomacy. Bringing people around the world together to cook and share food.

Planetary Citizens
325 Ninth Street, San Francisco, CA 94103
(415) 255-7008

One-world consciousness, work with U.N.; journal, newsletter, planetary passport, tapes, resource information.

PLENTY
PO Box 2306, Davis, CA
(916) 753-0731

International development projects; community organizing to relieve hunger and poverty; agricultural information, ambulance service in the Bronx; newsletter.

Rainforest Action Network
301 Broadway, Suite A, San Francisco, CA 94133

Part of an international network fighting to save rain forests, RAN employs tactics ranging from boycotts and protests to scientific symposiums and media campaigns.

Rocky Mountain Institute
1739 Snowmass Creek Rd., Snowmass, CO 81654-9199
(303) 927-3128

A mountaintop think tank established by renewable-energy gurus Hunter and Amory Lovins. Responsible for ground-breaking work on the boomerang effect of economic development plans in small towns and the use of energy self-reliance as a better means of national security than military expenditures.

Saybrook Institute
1550 Sutter St., San Francisco, CA 94109
(415) 441-5034

Center for advanced education in psychology, communications and global awareness, US/USSR citizen diplomacy involvement.

Sea Shepherd Conservation Society
Box 7000-S, Redondo Beach, CA 90277
(213) 394-3198

Focuses on the protection of marine mammals and habitats. Through public education, research, and physical confrontation with those who exploit marine wildlife, the organization has stopped more killings than any other society and hopes to end killing of whales, dolphins, and seals throughout the world. Publishes a quarterly newsletter.

Search for Common Ground
2005 Massachusetts Ave., NW, Washington, DC 20036
(202) 265-4300

Effective organization for bringing together leaders of opposing factions and beliefs to discover "common ground."

Seva Foundation
108 Spring Lake Drive, Chelsea, MI 48118
(313) 475-1351

An international service organization of caring people dedicated to relieving suffering. Seva is the Sanskrit word for service. The Seva family comprises supporters from 36 countries with many different philosophical, religious and political views. All are bound together by a common vision of a world in which serving others is the most precious product of our lives.

Sierra Club
730 Polk St., San Francisco, CA 94109
(415) 923-5660

A non-profit, member-supported organization that promotes conservation of the natural environment by influencing public policy decisions — legislative, administrative, legal, and electoral. Active with education and has many publications.

Sirius Community
Baker Road, Shutesbury, MA 01072
(413) 259-1251

Spiritual community and educational center. Books, retreats, education, workshops, tapes, slide shows.

Starcross Monastery
PO Box 14279, Santa Rosa, CA 95402
(707) 886-5446

Non-denominational monastery, retreat center, and home for disturbed and unwanted children and infants with the AIDS virus.

Wilderness Society
1400 I St. NW, 10th Floor, Washington, DC 20005
(202) 842-3400

Works to ensure that all decisions made about the preservation, use, and management of public lands are made in the public interest. Promotes citizen action on issues affecting public lands; testifies at congressional hearings; sponsors conferences, seminars, and workshops on public-land management; supports public education projects; and publishes handbooks, studies, and the quarterly journal *Wilderness.*

Windstar
2317 Snowmass Creek Rd., Snowmass, CO 81654
(303) 937-4777

Conference center for the investigation of major planetary issues.

World Watch Institute
1776 Massachusetts Ave. NW, Washington, DC 20036-1904
(202) 452-1999

A respected non-profit think tank that takes a comprehensive look at global environmental, economic, and social issues. Publishes a good magazine, informative special reports and *State of the World*, an annual report that *Time* magazine said was more influential around the world than Reagan's State of the Union messages. World Watch's style is conservative, but its recommendations often are not.

Zero Population Growth
1400 16th St. NW, Suite 320, Washington, DC 20036
(202) 332-2200

Works for a sustainable balance of people, resources, and environment. Thirteen chapters of local volunteers. Members call, write, and visit congressional representatives, assist in teacher-training programs, and monitor media coverage of population issues.

Earthstewards Network

Earthstewards have committed themselves to the spreading of consciousness, based upon the Sevenfold Path of Peace:

> When we are at peace within our own hearts we shall be at peace with everyone and with our Mother the Earth.
>
> When we recognize that our planet itself is a living organism co-evolving with humankind we shall become worthy of stewardship.
>
> When we see ourselves as stewards of our planet and not as owners and masters of it there shall be lasting satisfaction from our labors.
>
> When we accept the concept of Right Livelihood as the basic right of all we shall have respect for one another.
>
> When we respect the sacredness of all life we shall be truly free.
>
> When we free ourselves from our attachment to our ego-personalities we shall be able to experience our Oneness.
>
> When we experience our Oneness—our total connectedness with all beings—we shall be at peace within our own hearts.

As an Earthsteward, you will receive materials from many sources and be part of a network which will assist you in turning the thrust of humanity toward more holistic, loving, sharing relationships with each other and with all life forms, including our planet itself.

A commitment that you will make as an Earthsteward is to develop a program of action which relates the concepts of the Sevenfold Path of Peace to your daily life. This commitment is to yourself and only you define what it shall be. Material is provided to assist you in your planning, but you alone create the nature of the commitment. In this way, Earthstewards everywhere are literally changing the world. Our present world of aggression, excess, and isolation will not be changed by massive countermovements; it will be transformed by you and me and thousands like us who make a commitment, who take on a sacred obligation to make a difference in whatever ways we can, with the tools we now have. Don't be worried about identifying what you can do. The function of the Earthstewards Network is to assist you in discovering your own path of service, and to act as a wonderful support group to call upon.

It is also suggested that you, as an Earthsteward, practice tithing to any groups organizations, or projects that you consider worthy of your support. Tithing is an ancient, well-proven concept; the number ten (tithe) is significant. Many Earthstewards allocate 10% of some segment of their financial resources (10% of gross income or net income or "pocket" money, etc.) to worthwhile efforts. Casting your bread upon the waters is a powerful way of linking your consciousness with that of all consciousness on our planet. It is important that you as an Earthsteward take responsibility for the expenditure of your money as well as your time and energy. No one will ask you about your tithing commitment; this is an agreement with yourself.

If you are ready to make a commitment, then you are ready to take your place in the Earthstewards Network. Please complete the application form so we can send information and materials to help you in creating your plan of action. You and the world stand on the brink of the next level of consciousness. For your own transformation–for our planet's transformation—take the leap. Become an Earthsteward!

Earthstewards Network Application

Name ____

Address ____

City ______________________________ State _____ Zip _________

Phone ( ) ______________________________ Date ____________

Please take a few minutes to list three of your favorite interests/skills/passions/resources. (No more than 25 characters per line, please.)
May we print your name address, telephone and interests in our annual Earthstewards Network Directory? ☐ Yes ☐ No

1. ____

2. ____

3. ____

When you become an Earthsteward, you will receive materials from many sources and be a part of an exciting network of people who support one another as they co-create a more peaceful, caring world. You will be invited to participate in all trainings, trips and gatherings, including the annual network-wide Earthstewards Gathering where hundreds of us come together to joyfully learn and laugh and connect. Your initial Earthstewards packet contains the Earthstewards Handbook, your personal commitment chart (to help you design your own path of action and service), first-person stories from Earthstewards who want to share their experiences, and world-wide networking information. You will receive the Earthstewards Network Newsletter about every three months and regular informational updates concerning projects and programs.

☐ I am enclosing a check for $25 and wish to be included in the Earthsteward's Network. Outside the U.S., dues are U.S.$35, payable by International Money Order.

The Earthstewards Network
PO Box 10697, Bainbridge Island, WA 98110 206-842-7986

About the Authors

DANAAN PARRY is a psychologist and nuclear physicist, and the director of the Earthstewards Network, a nonprofit educational organization which focuses on conflict resolution, citizen diplomacy and global communication. The Earthstewards Network is an active global network of over 1500 individuals who are committed to positive, peace-filled change on our planet, and the awareness of the connection between all life.

LILA FOREST is a minister in both the Unitarian Universalist and Sufi traditions. In 1980, after founding and living in a spiritual community in the redwoods of northern California for 5½ years, Lila and Danaan Parry founded the Earthstewards Network to "bring the wisdom down from the mountain," and to apply it to daily life. Now she travels around North and Central America, writing and ministering to individuals and communities.

Sunstone Publications

We hope you have enjoyed *The Earthsteward's Handbook* as we are glad to have been able to share it with you. If you would like another copy for a friend or would like to purchase one of our other books, they are described below:

The Earthsteward's Handbook is a collection of reflections, ideas, inspiration and practical suggestions for making the spiritual vision of peace a reality in our life together on the planet. *Danaan Parry and Lila Forest.* ($6.95)

Warriors of the Heart is a handbook for bringing harmony to all our relationships, personal and planetary, based on conflict resolution techniques used by the author throughout the world. 224 pages. *Danaan Parry.* ($9.95)

The Essene Book of Days is a daily journal, calendar and guide for those on a path of personal and spiritual growth. Each daily page has an opening ritual, meditation, blessing, focus for the day, moon phase and sign, sun's light and sign and a place to record feelings. Also contains several teaching stories, information about the Essenes and the holidays surrounding the solstices, equinoxes and cross-quarter days. 416 pages. *Danaan Parry.* ($12.95)

Astro-Dome® 3D Map of the Night Sky contains everything you need to construct a 20" planetarium. Stars glow in the dark and a 24-page Constellation Handbook is included. *Klaus Hünig.* ($9.95)

Astro-Dots is a unique coloring book in which children connect the dots and stars to create the constellations. Fun and educational. ($3.95)

To Order: Include $2.50 for shipping and handling the first item, and 50¢ for each additional item. Send check, money order or Visa/Mastercard number to: Sunstone Publications, PO Box 788E, Cooperstown, NY 13326. Or phone 800-327-0306. NY residents please add appropriate sales tax. Write for our free catalog.

. . . In loving it is done

Navajo prayer